GUIDED WORKSHEETS

THINKING QUANTITATIVELY
COMMUNICATING WITH
NUMBERS

UPDATE

Eric Gaze

Bowdoin College

PEARSON

Boston   Columbus   Indianapolis   New York   San Francisco

Amsterdam   Cape Town   Dubai   London   Madrid   Milan   Munich   Paris   Montreal   Toronto

Delhi   Mexico City   São Paulo   Sydney   Hong Kong   Seoul   Singapore   Taipei   Tokyo

www.pearsonhighered.com

A Note from the Author to the Student

The purpose of this Quantitative Reasoning (QR) course is to provide a comprehensive overview of the quantitative skills required to cope with the practical demands of daily life, as well as preparing you for a deeper understanding of information presented in mathematical terms. Critical thinking and problem solving are an emphasis. The application of quantitative skills to decision making, requiring reasoning from evidence, will enhance your personal, civic, and business lives. These reasoning capabilities are based upon the ability to communicate with numbers effectively, so developing your quantitative literacy is a key focus.

What Is Quantitative Reasoning?

Quantitative Reasoning is the **skill set** necessary to process quantitative information and the **capacity** to critique, reflect upon, and apply such information in making decisions. Also called *numeracy*, it encompasses not just mathematical ability but also a disposition to engage quantitative information in a reflective and systematic way and use it to support valid inferences.[1] Derek Bok, in his book *Our Underachieving Colleges* (p. 68), provides a useful list of qualities of mind and habits of thought related to critical thinking that we will attempt to cultivate in this course.

The ability to:

- Recognize and define problems clearly
- Ask pertinent questions
- Identify arguments/issues on all sides
- Gather relevant facts . . . appreciate their relevance
- Perceive as many plausible solutions as possible
- Exercise good judgment in choosing solutions
- Use inference/analogy/logic to test the cogency of arguments

---

[1]Peters, E., Västfjäll, D., Slovic, P., Mertz, C. K., Mazzocco, K., & Dickert, S. Numeracy and decision making. *Psychological Science* 17, 407–413 (2006).

The Approach

The quantitative skill set will be built around the unifying theme of ratios and proportional reasoning, providing you with a coherent framework to connect the different topics covered. Ratios and proportions are mathematical concepts you learned starting in middle school. We will break down problems that appear complex and see how at their root they are based on the familiar concepts of ratio and proportion. This should help relieve any mathematical anxiety you may have about this course, and instill in you a confidence in approaching problems that are quantitative in nature. Recognizing that most problems encountered in daily life require just a few fundamental solution techniques rooted in proportional reasoning will help you see problems in a new light. We will use the spreadsheet program Excel to help develop your quantitative reasoning abilities, thus providing you with valuable computer skills to carry you throughout your college career and personal and professional lives. No more wondering "when am I ever going to use this?" that is so common in traditional math courses! The utility of the mathematical concepts encountered in this course are front and center, with applications providing context for all of the content.

The input/output interface of Excel is a terrific way to deepen your algebraic reasoning skills as you explore the relationships between quantities. Spreadsheets are built to work with data, so Excel will facilitate our introduction of statistics in the *Spotlight on Statistics* sections found at the end of most chapters. In addition, financial applications are most easily studied using the power of spreadsheets, so your financial literacy will be systematically developed through applications and examples throughout the course.

The text *Mathematics and Democracy*[2] gives the following justification for mastering the above-mentioned skills:

 Quantitatively literate citizens need to **know more than formulas and equations**. They need a predisposition to look at the world through mathematical eyes, to see the benefits (and risks) of thinking quantitatively about commonplace issues, and to **approach complex problems with confidence** in the value of careful reasoning. Quantitative literacy **empowers people** by giving them tools to **think for themselves**, to ask intelligent questions of experts, and to confront authority confidently. These are **skills required to thrive in the modern world**.

The Course

Why should you take this course? To "thrive in the modern world" requires a fundamental literacy with quantitative information. One hundred years ago it was critical to be able to read

[2]http://www.maa.org/ql/001-22.pdf

and write to participate fully in all aspects of our society. Today the same holds true for being quantitatively literate. It is not just that many careers are requiring more and more quantitative reasoning skills, as our economy becomes increasingly driven by data and computation; but all aspects of your life require more QR, from personal finance, to health-related decisions, to voting on public policy issues. Data truly exist everywhere in the 21st century, and organization and analysis of data start with a basic facility with spreadsheets. You will leave this course confident in your ability to use Excel and in your ability to have meaningful discussions involving quantitative information.

This course will probably be unlike any other math course you've taken. You will not be asked to plug and chug numbers. You will not be asked to factor a quadratic polynomial. You will not be asked to memorize formulas. You will not be asked to do rote, boring homework. You will be asked to look at problems from a new perspective. You will be asked to think and question and think again. You will be asked to think deeply about quantitative problems, and share your insights through effective communication. If my students are any indication, this will be one of the most rewarding and beneficial math classes you've taken. Throw out your preconceived notions of a math class and get ready to really dig in, explore, develop your reasoning ability, and have fun!

The single biggest obstacle to success in anything (not just this math course) is how willing you are to participate and persist in the endeavor. Your mere presence in the classroom, while necessary, is far from sufficient for success (the plants in my office don't seem to have learned much over the years). You must learn to be Present with a capital P. Boredom, dislike, and fear are all states of mind that can be changed! The following five states of mind (AEIOU) should be cultivated, and all negative thoughts about this course and its possible outcomes should be banished. To get the most out of this course requires full participation. Your overall enjoyment and success are not dependent on your instructor but on how you engage with the material.

1. **Be Active** in all parts of the learning process from the classroom to homework.

2. **Be Engaged** in what others are saying and in what you are doing.

3. **Be Interested**, as it is so much more fun than being bored.

4. **Be Observant** and reflect on what you see, and ask questions as much as possible.

5. **Be Uplifting** toward others and yourself.

Resources

Guided Worksheets
ISBN-10: 0-13-454044-1   ISBN-13: 978-0-13-454044-3
The Guided Worksheets are a key element to truly grasping the content. These worksheets are downloadable within the course or available as a printed booklet. If your instructor did not

order the Guided Worksheet booklet and you'd rather have the booklet than download the pages, you can purchase it from the Purchase Options tab in the left-hand navigation bar of your course. Use the worksheets to take notes, jot down questions or ideas, and work through the example problems. Class sessions will be more interesting if you take full advantage of the Guided Worksheets.

Flashcards (online only)

In Tools for Success you'll find a link to an engaging flashcard application to help you learn the vocabulary from the course. Some chapters do have a lot of vocabulary, and the better and sooner you learn it, the easier it will be to understand the explanations.

Thinking Quantitatively blog

http://thinkingquantitativelystudents.wordpress.com/

I thoroughly enjoy talking about quantitative reasoning and aim to create a forum in which you can participate in the discussion. Quantitative reasoning involves thinking about real life issues so we can share news and discuss the findings in news reports. You can share what you're doing in your class and learn what students at other colleges are doing in their courses. I think you're going to really enjoy this course, and the *Thinking Quantitatively* blog will be a place where you can share that enthusiasm.

RSS Feeds

In Tools for Success you'll see an RSS Feed link. This is organized by topics covered in the course and will provide news on a daily basis. Use these feeds for ideas for projects or test yourself to analyze articles and real data to see how much you've learned and how you've begun to think differently!

Some Practical Advice!

Here is some final practical advice on how to go about succeeding in this course. Critical thinking requires a certain tolerance for ambiguity that some students are uncomfortable with, especially in a math class. When confronted with a real world problem, often it is not clear what action to take or what relevant information is needed for the solution. The following three questions can magically clear away the "fog of ambiguity" and set you on a constructive path to the solution. Therefore, don't despair if you ever experience a problem in which your first reaction is, "What in the world is going on?" Ask these three questions instead!

1. What information is given?

2. What are they looking for?

3. How can I use the given information to find this?

Let's use an example from the course to illustrate the power of these three magic questions.

Example: Your car averages 27 miles per gallon (mpg), and you drive approximately 15,000 miles in a year. Assuming the cost of gas is $3.40 a gallon, how much can you anticipate spending on gas in a given year?

1. What information is given?

    a. 27 mpg

    b. 15,000 miles in a year

    c. $3.40 per gallon

2. What are they looking for?

    a. They are asking for the total money spent on gas in a year, so a dollar amount.

3. How can I use the given information to find this?

    a. They are looking for a dollar amount, so we need to use the $3.40 per gallon. If we know the number of gallons used in a year we can multiply $3.40 by the number of gallons to get the total amount spent on gas (total expenditure):

$$\text{Cost(\$)} = \frac{\$3.40}{1 \text{ gal}} \cdot x \text{ gal}$$

Alternatively, we could think of this as a proportion:

$$\frac{\text{Cost (\$)}}{x \text{ gal}} = \frac{\$3.40}{1 \text{ gal}}$$

    b. We are not given the number of gallons used in a year, but we are given the number of miles driven in a year: 15,000. Maybe we can use this somehow to find gallons.

    c. We have not used the 27 mpg yet. This is telling us we can drive 27 miles per 1 gallon of gas. So we use 2 gallons for 54 miles and 10 gallons for 270 miles, hmmm . . . maybe we can set up a proportion using 27 miles per 1 gallon and the 15,000 miles:

$$\frac{27 \text{ miles}}{1 \text{ gal}} = \frac{15,000 \text{ miles}}{x \text{ gal}}$$

Cross multiplying gives us:

$$27 \cdot x = 1 \cdot 15,000$$

$$x = \frac{15,000}{27} = 555.56 \text{ gal}$$

d. Great! Now we know the gallons used in a year, so we can plug this number into the equation from part **a** (or use the proportion), and solve for cost:

$$\text{Cost (\$)} = \frac{\$3.40}{1 \text{ gal}} \cdot 555.56 \text{ gal} = \$1{,}888.90$$

That's a lot of money! In addition to maintenance and insurance, the additional cost of gas makes owning a car expensive. This is the sort of calculation that a quantitatively literate person is comfortable carrying out when confronted with a real world decision like buying a car.

I hope this example illustrates what is meant by quantitative reasoning, and how proportional reasoning (setting up a proportion and solving) underlies much of the mathematics we carry out on a day-to-day basis. The three questions provide concrete action items to work on, so we don't get stuck wallowing in a fog of numbers. This course will teach you how to confidently carry out calculations such as the ones in this example, and empower you to embrace numbers in all aspects of your life.

Best of luck!

Eric Gaze

Table of Contents

Quantitative Literacy

A piece presented on the Bloomberg View explores the data regarding "How Americans Die." Let's take a closer look.

1. On the first slide use the statistics for 1968: 823.7, 967.3, and 1,118.5, in sentences.

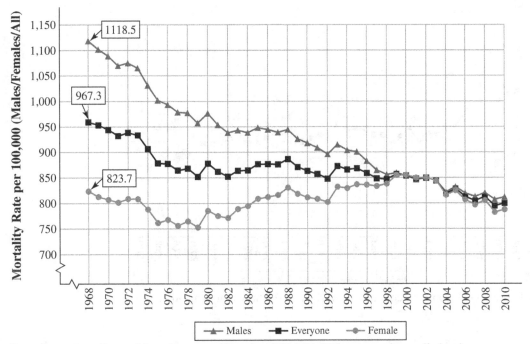

Data from: http://www.bloomberg.com/dataview/2014-04-17/how-americans-die.html

2. The presentation tells us the overall rate "fell by about 17%" from 1968 to 2010, from 967.3 to 799.5.

    a. Verify this and quantify the change for the rates for men and women in a similar fashion.

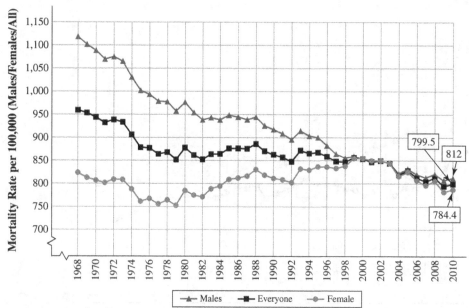

    b. Why can we compare the 1970 and 2010 statistics, even though the population has increased over this period?

3. Slide 1 says the decline in mortality rates stops in the mid 1990's and slide 2 attributes this to the aging of the population.

    a. What is the logic behind this argument?

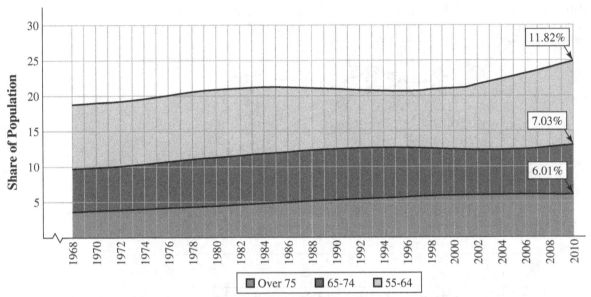

Data from: http://www.bloomberg.com/dataview/2014-04-17/how-americans-die.html

    b. Interpret the 11.82% for 2010 in slide 2 and compare to the 25% on the vertical axis.

4. Looking at slide 4 which line stands out from the rest? What do you think accounts for this difference?

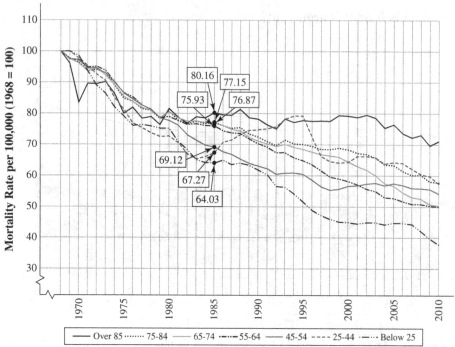

Data from: http://www.bloomberg.com/dataview/2014-04-17/how-americans-die.html

5. Interpret the statistics 80.16 and 64.03 for 1985 on slide 4. Hint: Compare to slide 3 statistics shown here.

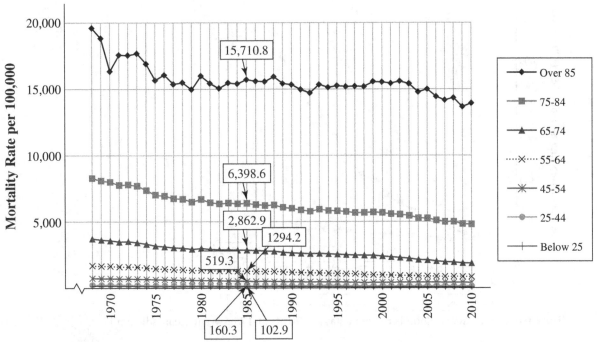

Data from: http://www.bloomberg.com/dataview/2014-04-17/how-americans-die.html

6. Deaths from drugs in the 45 to 54 year old populations have increased by what factor from 1990 to 2010 as shown in slide 11?

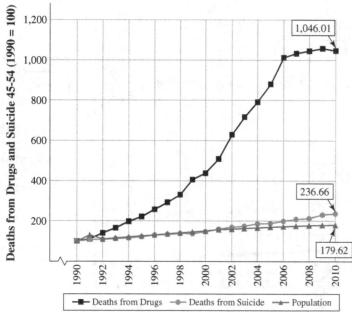

Data from: http://www.bloomberg.com/dataview/2014-04-17/how-americans-die.html

7. Does slide 17 indicate Medicare spending has been increasing or decreasing since 2010?

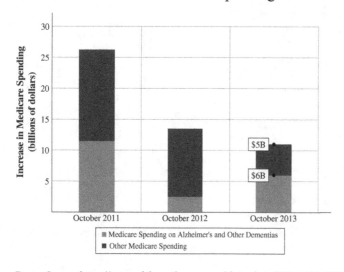

Data from: http://www.bloomberg.com/dataview/2014-04-17/how-americans-die.html

Quantitative Reasoning / Introduction to Excel

Why go to college? What is the PURPOSE of a college education? List 3 specific purposes:

What is critical thinking? List 3 *characteristics* of critical thinking:

Introduction to Excel

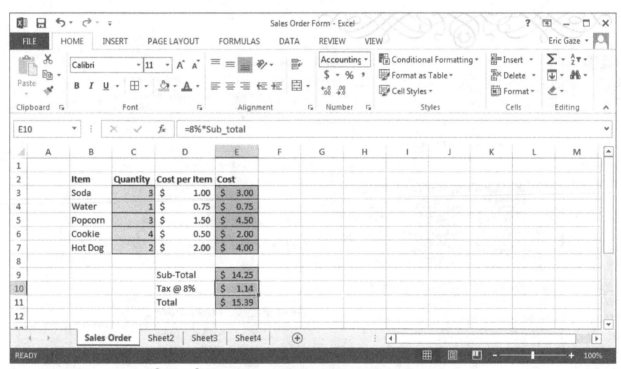

Screenshots from Microsoft® Excel®. Used by permission of Microsoft Corporation.

1. What formula is entered in cell E3?

2. How do you fill this formula down?

3. What happens to the cell references in the formula when you fill down?

4. What built-in function can be used in cell E9? What formula using this function is in cell E9?

5. If you format the Tax in cell E10 to show zero decimal places what happens to the output in cell E11?

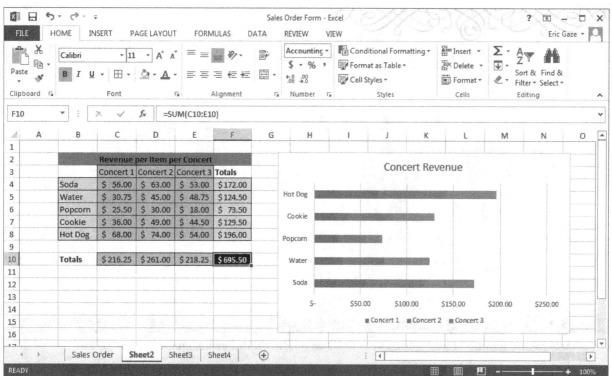

Screenshots from Microsoft® Excel®. Used by permission of Microsoft Corporation.

6. How do you change the name of the sheet tabs?

7. What button do you hit in the menu to format the numbers as currency? What button makes borders?

8. To create the chart what cell range was highlighted in the worksheet?

9. What type of chart is this?

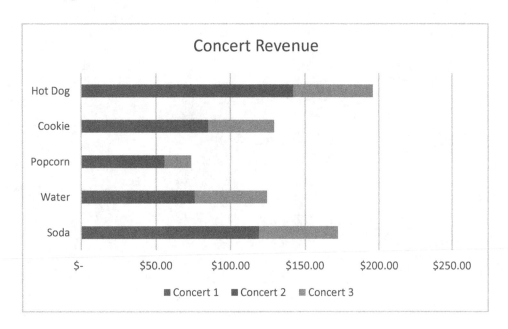

Play with Excel and create some more charts. Have Fun!

Function and TV Loan

> **Definition:** A **function** is a relationship between quantities referred to as **inputs** and **outputs**, in which every collection of inputs is paired up with one and only one output.

1. Determine which of the following are functions:

INPUTS	OUTPUT	FUNCTION?
States	Senators	
Senators	States	
States	Number of senators	
People	Anyone they been married to…	
People	Number of spouses	
US Citizens	Social Security Numbers	
SS#'s	US Citizens	
People	Birthdays	
Birthdays	People born on that day	
Students in this class	Shoe size	
Shoe Size	Students in this class with that shoe size	

2. Come up with a function between two quantities which remains a function when you switch the inputs and outputs…

Next we will explore some functions associated with taking out a loan.

Financial Literacy Vocabulary for Loans

- *Principal*: The amount of money borrowed from the lender.

- *Interest*: The money or fee the lender charges you for borrowing money.

- *Period*: The length of time before your next payment is due and interest is charged; typically 1 month for most loans.

- *Balance*: What you owe at the end of each period factoring in any interest and payments made.

- *Interest Rate*: The ratio of interest charged to amount owed, typically represented as a percentage which is a rate per 100. An interest rate of 6% means you will be charged $6 for every $100 you owe. Ratios will be covered in Chapter 2, rates in Chapter 3 and percentages in Chapter 4.

- *Annual Percentage Rate (APR)*: The interest rate for a period of 1 year.

- *Periodic Rate*: The interest rate for a period other than 1 year, it is the APR divided by the number of periods in a year: *APR/n*. A 6% APR computed monthly will give a 6%/12 = 0.5% periodic rate.

- *Annual Percentage Yield (APY)*: Given a periodic rate, your interest will *compound*. The APY is the ratio of the total interest charged for the year to the original principal.

Credit Card Loan

Let us assume you buy a Sony flat screen TV for your dorm room that costs $1,000. You make the purchase with a store credit card that has a 12% *Annual Percentage Rate* (APR). You do not have to make any monthly payments for the first year (sounds good!), but they will charge interest at the end of each month. How much do you owe at the end of the first year?

3. What is the Principal?

4. What is the monthly interest rate?

5. What is the interest charged for the first month?

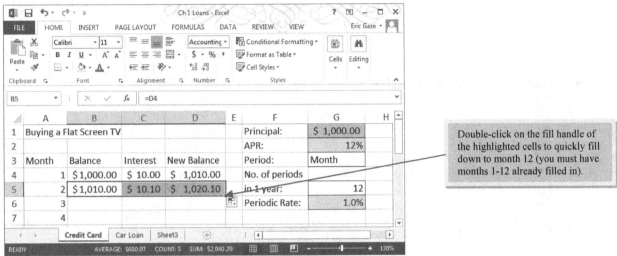

Screenshots from Microsoft® Excel®. Used by permission of Microsoft Corporation.

Cell Reference Handbook

Relative	**A4**	**Changes** the row number when you fill up/down and **changes** the column letter when you fill left/right.
Mixed	**$A4**	**Changes** the row number when you fill up/down and *fixes* the column letter when you fill left/right.
Mixed	**A$4**	*Fixes* the row number when you fill up/down and **changes** the column letter when you fill left/right.
Absolute	**A4**	*Fixes* the row number when you fill up/down and *fixes* the column letter when you fill left/right.

6. What formula is in cell C4? D4?

7. Which formula involves an *absolute cell reference*?

8. Why do we need to fill in the second row before we fill the formulas down?

9. What do you owe at the end of the first year?

10. What is the APY?

Car Loan

You have just bought a new VW Jetta for $17,254.38. You put down $2,254.38 of your savings so you only have to borrow $15,000. The auto dealership gets you a loan from a bank for 5 years at 6%, which you agree to and sign. What will be your fixed monthly payment?

The periodic payment is a function of 4 inputs:

INPUTS	OUTPUT
Principal (P)	Periodic payment (PMT)
APR	
Number of periods in 1 year (n)	
Number of years of the loan (t)	

$$PMT = \frac{P \times \dfrac{APR}{n}}{\left(1 - \left(1 + \dfrac{APR}{n}\right)^{-nt}\right)}$$

We are going to create the following spreadsheet:

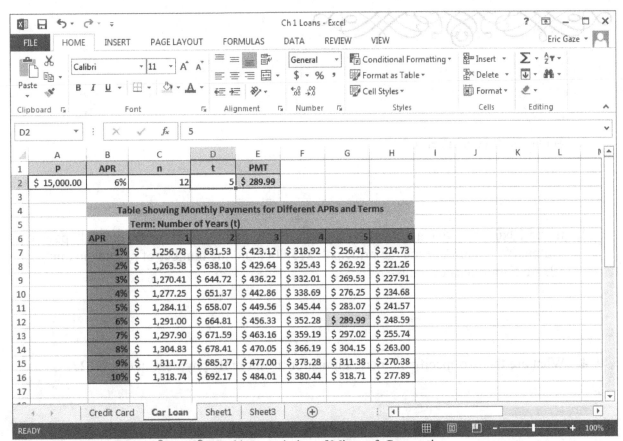

Screenshots from Microsoft® Excel®. Used by permission of Microsoft Corporation.

Caution! When entering formulas into Excel you must use **Order of Operations**:

1. *Parentheses:* Everything entered in parentheses will be computed first. When in doubt use parentheses, especially for the numerator and denominator of fractions. Parentheses are like Vitamin C, too much won't hurt you (you just have to use the rest room a lot) but too little and your formula will get gangrene and rot.

2. *Exponents:* Exponents are next, use the $^\wedge$ symbol above the number 6. Complicated exponents need parentheses: $2^{\frac{1}{3}} = 2^\wedge(1/3)$.

3. *Multiplication and Division:* These are tied, Excel will compute from left to right.

4. *Addition and Subtraction:* These are also tied and will be computed from left to right.

5. *PEMDAS:* Please Excuse My Dear Aunt Sally is the traditional mnemonic device to help remember the order of operations.

6. *Examples:*
   a. $= 3 * 5 - 2 = 13$
   b. $= 3 * (5 - 2) = 9$
   c. $= 5 + 2 * 3 = 11$ (not 21)

1. Evaluate $=2-6/4+2$

2. What formula is in cell **E2**?

3. In the 10x6 table what are the two inputs (variables) for each of the 60 monthly payments (outputs)? Note: Assume the principal, $15,000, and the number of months in one year, 12, are both fixed constants.

4. What are the two specific numeric inputs for the monthly payment in cell **G12**?

5. What are the two specific numeric inputs, and associated cell references, for the formula in cell **C7**? Which of these should vary as we fill the formula down?

6. What formula is in cell **C7**?

7. How much total interest do you pay with the $289.99 monthly payment?

8. How much would you save if you switched from the 5 year 6% loan to a 5% APR?

9. How much would you save if you switched from the 5 year 6% loan to a 3 year 6% loan?

10. If you have horrible credit you might be forced to take out a 6 year 10% loan. How much interest do you pay over the life of this loan?

11. What is fixed (column or row) by the following cell references?

F4	G$2	T1	$V85	J$21	$S99

Descriptive Statistics

We will use the *CAT scores* sheet in *Data Sets:*

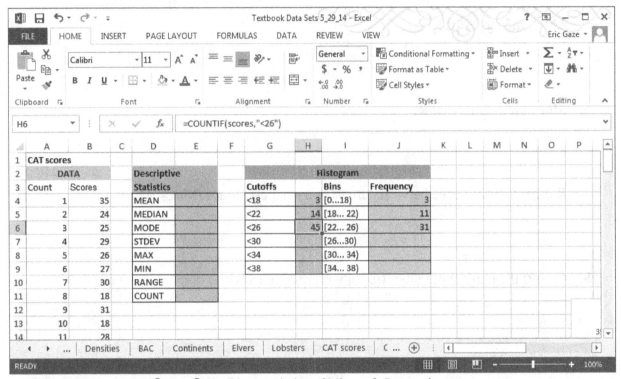

Screenshots from Microsoft® Excel®. Used by permission of Microsoft Corporation.

Descriptive Statistics: Note that the name of each function in Excel is as given in the spreadsheet, except for the mean which uses the **AVERAGE** function, and the range which is **MAX - MIN**.

Enter the appropriate functions into the spreadsheet to compute the following, and write the formula you would type into Excel after each definition (note there are 98 scores listed in column **B** and the cell range **B4:B101** has been named *scores*).

1. *Mean*: the arithmetic average.

2. *Median*: the middle of the data set (half above, half below, 50th percentile).

3. *Mode*: the most frequently occurring value.

4. *Standard Deviation*: the "average distance" of the data values from the mean.

5. *Max*: the largest value.

6. *Min*: the smallest value.

7. *Range*: the difference between largest and smallest values (Max – Min).

8. *Count*: the number of values (usually referred to as N).

Histogram

Note the formula bar in the spreadsheet giving the function =**COUNTIF**(scores,"<26").

9. What is the output 45 telling us in cell **H6**?

10. How do you *name* a cell range *scores*?

11. The **COUNTIF** function has two arguments, what are they called in general?

12. What formula should be typed into cell **H7**?

13. Why is there 14 in cell **H5** but only 11 in cell **J5**?

14. What formula is in cell **J6**?

Finish entering formulas in the Histogram box and create a column chart of the last two columns, *Bins* and *Frequency*. This is called a Histogram.

Complete the histogram by drawing in the columns for the last three bins.

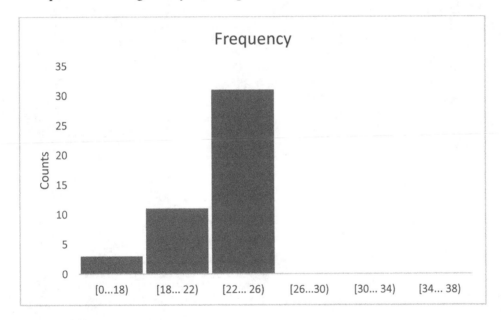

Ratios!

> **Definition:** A **ratio** is a comparison of the *relative* size of two or more quantities. If the ratio of quantity A to quantity B is 2 to 5, or 2 : 5, then for every 2 units of quantity A there are 5 units of quantity B.

What are the two quantities being compared in the following ratios?

Ratio	Thing 1	Thing 2
GPA is 2.74		
Only 1 U.S. Adult in 5 could calculate interest		
13% of U.S. Adults are numerate		
APR = 6%		
Speed is 7.24 mph		
Half of 17 year olds don't know enough math to work in auto plant		
Math SAT scores for Elem. Ed. majors compared to Natl. Avg. is 483 vs. 515		
You borrow $300 and pay $350 at the end of the month		
52% of freshman say their emotional health is above average		

Ratio Tables are a nice way to organize quantitative relationships:

Tuition and Fees 2013-14

Private 4-year	Public 4-year (In-State)	Public 2-year
$30,094	$8,893	$3,264
	1	
100		
		1

Multiply by ... ?

We have scaled this quantity to 1.

1. Determine the ratio of Private 4-year to Public 4-year In-State with the second quantity scaled to 1 using the fundamental solution technique to compute ratios:

    1. Set up proportion:

    2. Cross multiply and solve!

2. Fill in the values of all the boxes in the ratio table. For the greatest accuracy you should always try to use the original statistics in the 2nd row of the table.

Tuition and Fees 2013-14

Private 4-year	Public 4-year (In-State)	Public 2-year
$30,094	$8,893	$3,264
	1	
100		
		1

3. What number do you multiply by to go from the values in the 3rd row to the 4th row? From 3rd column to the 1st column? Note that there will be some discrepancy with the values obtained by doing this and those in the table due to rounding.

Number of courses taken during freshman and sophomore year
(standardized cohort of 1,000 students)

Division and Department	Whites	Asians	Latinos	Blacks
Humanities				
Art and Art History	541	459	505	291
Classics	84	52	42	51
Comparitive Literature	55	57	102	99
English	1,567	1,414	1,536	1,589
Foreign Language	1,685	1,517	1,805	1,714
Spanish	*593*	*337*	*918*	*807*
Other	*1,092*	*1,180*	*887*	*990*
History	827	584	811	751
Music	782	823	655	657
Philosophy	519	426	446	406
Religion	357	249	288	248
Rhetoric	22	18	16	21
Theater and Media Arts	298	160	306	278
Other Humanities	768	357	893	801
Total	*7,505*	*6,116*	*7,405*	*6,906*
Ratio to Whites	*1.00*	*0.81*	*0.99*	*0.92*

Data from: Table 2.1 in the book, *Taming the River*, by Charles, Fisher, Mooney Massey; Princeton Press page 25

4. What does the number 541 mean in the first column?

5. Compute the ratio of English to Art classes for each column, and scale the ratios so that the second quantity is 1.

6. Compute the ratio of Art to English for each column, and scale the ratios so that the second quantity is 100.

7. How was the Ratio to Whites computed? What are the two quantities involved?

8. Create three ratios of your choosing and scale the second quantity to a nice number ☺

Weighted Averages $= \dfrac{weight_1 \times data_1 + weight_2 \times data_2 + \cdots + weight_n \times data_n}{sum\ of\ weights}$

1. You buy 8 children's tickets at $5.00 apiece and 2 adult tickets at $12.00 apiece.
   a. What is the average price per ticket?

   b. What are the *weights* in this weighted average problem?

2. Your grade consists of a test worth 80% and a paper worth 20%. You get a 70 on the test and a 100 on the paper.
   a. What grade do you get in the course?

   b. What are the *weights* in this problem?

3. Now assume there are two exams worth 40% each and a paper worth 20%. You get a 60 on the midterm and a 100 on the paper.
   a. What is your grade at this point in the course?

   b. What is the highest possible grade you can get in the class?

4. The following shows a gradebook for a class.

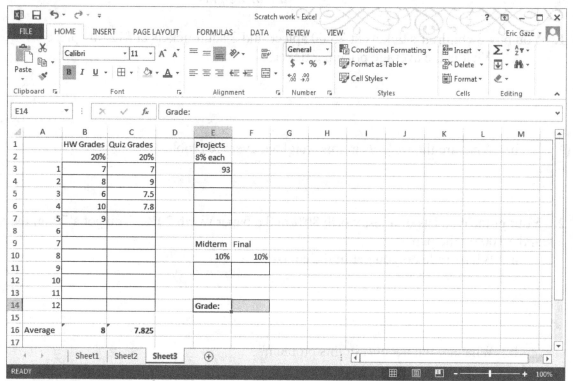

Screenshots from Microsoft® Excel®. Used by permission of Microsoft Corporation.

a. What formula is in cell **B16**?

b. What is this student's grade in the class right now?

c. What will her grade be if she gets a 100 on the next Project and a 95 on the Midterm?

Proportionality

- Determine which of the following quantities are proportional.
- If they are proportional, express their relationship using an equation.
- Specify the units of the constant of proportionality in your equation, and identify the name of the constant if you know it.

The first one has been done for you:

1. Property tax : Appraised value of your house

    - YES proportional

    - Tax = k * Value

    - $k = \dfrac{Tax\ \$}{House\ value\ \$} = \dfrac{\$}{\$}$ These units cancel, so this constant of proportionality is
    unit-less. It is called the *tax rate* or the *mill rate*.

2. population of a region : area of the region

3. volume of water : weight of water

4. population of NY City : population of NY state

Intuitive Definition: Two quantities are **directly proportional** if when you double one quantity …

_____

---

5. number of babies born in a year : number of women aged 12-45

6. the total cost of buying tickets : number of tickets

7. height of a digital photo : width of a digital photo

8. the number of accidental deaths in 2005 : the number of motor vehicle deaths

9. cups of sugar : cups of flour (in a pancake recipe)

10. your IQ : your GPA

11. distance to work : time it takes to get to work

12. distance to a celestial object : speed at which it is receding from us

13. euros : dollars (at an exchange booth)

14. miles you drive : gallons of gas used

CPI

The *Consumer Price Index* (**CPI**) is a financial ratio that has to do with inflation. The costs of goods and services typically rise each year. For example, a movie ticket in 1990 that cost $6.50 will cost more in 2005. Thus a dollar in 1990 (1990$) was worth **more** than a dollar in 2005 (2005$).

It is helpful to think of 1990$ and 2005$ as different currency in much the same way that dollars and euros are different currency. We need to convert between these currencies and the CPI allows us to do that.

The CPI is a measure of the cost of a fixed basket of goods and services in a given year. This is called an **index** because it has arbitrarily been set to 100 for the cost of this basket in 1983$. These same goods and services would have then cost 130.7 in 1990$ and 195.3 in 2005$.

To convert the cost of a $6.50 movie ticket in 1990 to 2005$ we can set up a proportion using the CPI values:

	CPI	COST OF AN ITEM
1990$	130.7	6.50
2005$	195.3	

1. Set up a proportion and compute the cost of the movie ticket in 2005$ which originally cost $6.50 in 1990$.

2. Assume the ratio of euros to dollars is 4€ : 5$. What is more expensive, a mocha latte that costs 2€ or one that costs $3.50?

		Cost of an Item	
	CONVERSION	2€ LATTE	$3.50 LATTE
EUR	4	2	
USD	5	x	3.5

3. Which was more expensive: a $2 Starbucks mocha latte in 1990 or a $3.50 Starbucks mocha latte in 2005?

Cost of an Item

	CPI	1990	2005
1990$	130.7	2	
2005$	195.3	*x*	3.5

The CPI allows us to compare datasets that involve monetary values over time. The amount of money the Federal Government collected (i.e. receipts (taxes)) looks to have been increasing steadily in the following table. But in order to compare these values we need to convert all three into **one currency**.

		1990	2000	2005
	CPI	US RECEIPTS (BILLIONS)		
1990$	130.7	1,032		
2000$	172.2		2,025	
2005$	195.3			2,154

4. Convert the 3 receipt values in the table to **one currency** using the CPI values (you can choose which currency to use), and determine the year when the receipts are greatest when adjusted for inflation.

		1990	2000	2005
	CPI	US RECEIPTS (BILLIONS)		
1990$	130.7	1,032		
2000$	172.2		2,025	
2005$	195.3	a	b	2,154

PE and Money Ratios

The Price to Earnings, or PE ratio, is one of the most common statistics used to assess the "value" of a share of stock. Stocks will be fully discussed in Chapter 9, but for now we want to introduce the basic idea of the PE ratio.

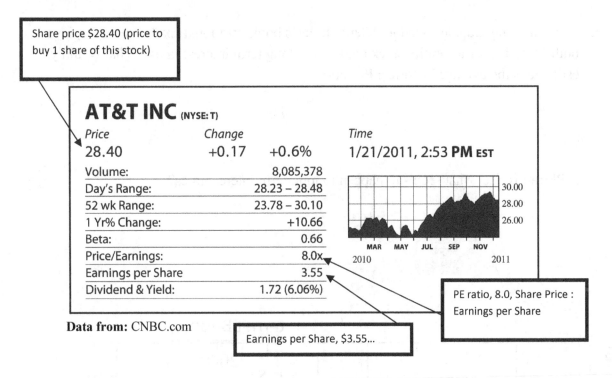

Share price $28.40 (price to buy 1 share of this stock)

AT&T INC (NYSE: T)

Price	Change		Time
28.40	+0.17	+0.6%	1/21/2011, 2:53 **PM** EST

Volume:	8,085,378
Day's Range:	28.23 – 28.48
52 wk Range:	23.78 – 30.10
1 Yr% Change:	+10.66
Beta:	0.66
Price/Earnings:	8.0x
Earnings per Share	3.55
Dividend & Yield:	1.72 (6.06%)

PE ratio, 8.0, Share Price : Earnings per Share

Earnings per Share, $3.55...

Data from: CNBC.com

1. Compute the ratio of the price per share to the earnings per share and scale the second quantity to 1.

2. The earnings per share (EPS) statistic is the ratio of the profits the company made for the previous 12 months to the number of the shares available. If there were 5.9 billion shares available, use the EPS statistic to compute the profits that AT&T made.

3. The share price changes constantly as people trade shares with each other. If there are more buyers than sellers, the price goes up. If there are more sellers than buyers the share price goes down. What are two different ways the PE ratio can increase?

4. Compute the ratio of the dividend to the price per share, and scale the second quantity to 100.

5. The following graphic found in Peter Schiller's book, Irrational Exuberance, displays both PE ratios for the entire stock market and long term interest rates. What would you estimate is the average historical PE ratio?

6. If PE ratios rise, what seems to happen to long term interest rates?

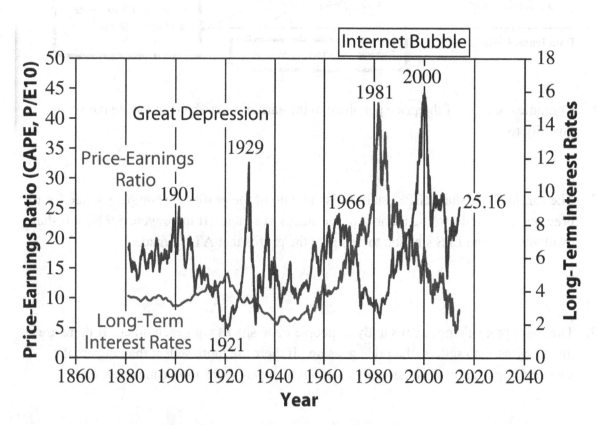

Figure 2.8 Schiller's Graphic, Showing Historical PE Ratios Against Interest Rates
Data from: *Irrational Exuberance* by Robert J. Shiller

Financial planning involves saving money for retirement and paying off loans for education and home ownership (mortgages). In this table you are given money ratios from the book, *Your Money Ratios*, comparing amounts related to your savings and loans. Income and earnings refer to your annual salary. These are just suggested guidelines meant to get you thinking about these topics, the stock : bond suggested allocation in particular varies widely depending on who you talk to.

Your Age	Captial : Income	Savings : Income	Mortgage : Income	Education : Earnings	Stock : Bond
25	0.1	12%	2.0	0.75	50% : 50%
30	0.6	12%	2.0	0.45	50% : 50%
35	1.4	12%	1.9	0.00	50% : 50%
40	2.4	12%	1.8	X	50% : 50%
45	3.7	15%	1.7	X	50% : 50%
50	5.2	15%	1.5	X	50% : 50%
55	7.1	15%	1.2	X	50% : 50%
60	9.4	15%	0.7	X	40% : 60%
65	12.0	15%	0.0	X	40% : 60%

Data from: Your Money Ratio$: 8 Simple Tools for Financial Security at Every Stage in Your Life by Charles Farrell

7. You are 35 years old and make $60,000. How much *Capital* should you have?

8. How much should you be saving this year?

9. How much should the balance on your mortgage be?

10. What should you owe on student loans?

11. How much of your capital should be invested in bonds?

Z-scores

You get a 63 on a test. **Is this bad?**

To answer this question we need to know how this score compares to the rest of the class. Assuming the mean was 51 we now know you scored above average. **But how far above average?**

To answer this question we need to know how spread out the scores are. If all the other scores are between 49 and 53 then you did extremely well (A+), but if a bunch of people got in the 90's and a bunch got in the 20's then you just did OK (B-).

The **standard deviation** (average distance from the mean) gives us a measure of this spread. Assuming it was 5 points, your 63 is more than twice that distance above the mean.

$$\textbf{Mean } \bar{X} = 51 \qquad \textbf{StDev } SD = 5$$

$$z = \frac{63 - 51}{5} = \frac{12}{5} = 2.4$$

In fact your 63 is 12 points above the mean and the distance of 12 is 2.4 times the average distance of 5 points. The number +2.4 is your **z-score**: it tells me you are 2.4 times the standard deviation from the mean. Someone who got a 39 is 12 points below the mean and would have a z-score of -2.4.

> **Definition:** The **z-score** of a data value is the ratio of the data value's distance from the mean to the average distance from the mean:
> $$z = \frac{x - \bar{X}}{SD}$$

1. If you get a 63 and the mean is $\bar{X} = 51$ and the standard deviation is $SD = 2$, what is your z-score?

2. Z-scores allow you to compare different scales. For example, the admissions office is comparing two different applicants, one has a math SAT score of 640 and the other has a math ACT score of 28. Which student has a better score? You need to know that for the SAT: $\bar{X} = 500$ and $SD = 100$, and for the ACT: $\bar{X} = 22$ and $SD = 5$.

3. The following table shows a dataset {69, 72, 75, 78, 81} with a mean of 75 and standard deviation of 4.74. Fill in the rest of the table using the formulas in the first column (some values have been computed for you):

	DATA VALUES					MEAN	ST DEV
Raw scores	69	72	75	78	81	75	4.74
Z-scores		-0.63			1.27	0	1
T-scores = 50+10*z					62.7	50	10
IQ = 100 + 15*z	80.95						
SAT = 500 + 100*z				563			

4. In the previous example with a mean of 75 and standard deviation of 4.74, what raw score would have a z-score of +1? -1? +2? -2?

5. A graduate school grades with the following system based on z-scores. If a class of 10 students takes a quiz and half get a 90 and half get a 92, the mean is 91 and the standard deviation is 1.05. What are their letter grades using this system?

A	$z > 2$
B	$1 \leq z < 2$
C	$z < 1$

6. Do you like or dislike this system? Explain!

Histogram and z-scores

Data Analysis: Multiple Representations

The grades in a class are {40, 60, 65, 72, 72, 75, 78, 78, 78, 83, 84, 87, 92, 95}. The mean $\bar{X} = 75.6$ and the standard deviation $SD = 14.1$. In this worksheet we are going to continue working with z-scores and explore how they relate to data distributions and histograms. Recall that:

$$z = \frac{x - \bar{X}}{SD}$$

1. In the following dot plot showing the distribution of grades, put tick marks on the x-axis from 40 to 100 every 5 units and indicate the mean with a tick mark.

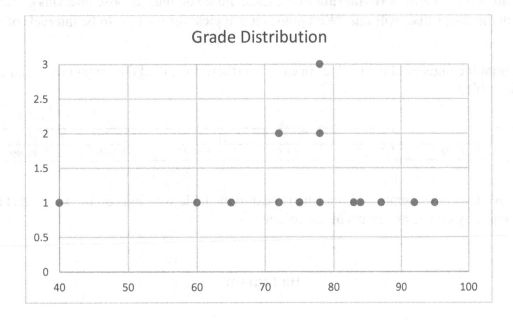

2. What x-value would have a z-score of +1? -1? -2? 0? Put a tick mark on the x-axis for each of these values.

x-value				
z-score	+1	-1	-2	0

3. We will treat the **0.5 axis** as the z-**score axis**. Using the dot plot for #1, put tick marks on the **0.5 axis** at $z = 0, \pm 1, -2$ and write these numbers below their respective tick marks.

4. Compute the IQ-scores, $IQ = 100 + 15 \cdot z$, and SAT scores, $SAT = 500 + 100 \cdot z$, for the data values in the table from question **#2**. The point here is that what matters is not the raw score or data value, but only the relationship of the data value to the center (mean) and the spread (standard deviation). Any of the following scores accurately represent this data set and could be used as the scale on the x-axis!

x-value				
z-score	+1	-1	-2	0
IQ-scores				
SAT scores				

Now we are going to draw a **Histogram** of the data. First you must choose intervals or "bins" into which the data values will fall. With grades it is typical for the bins to be intervals of length 10.

5. Count the number of data values in each bin (these are called the frequencies) and fill in the table.

Frequency						
Bins	40-49	50-59	60-69	70-79	80-89	90-99

6. Now draw a histogram (column chart) next to the table with bins on the x-axis and the frequency counts as heights of the columns.

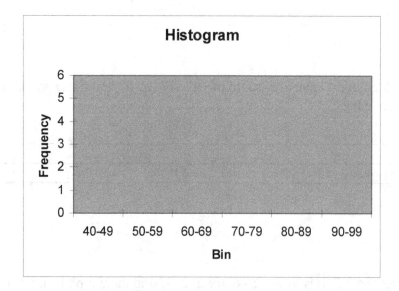

7. Why is the histogram referred to as a "distribution" of the data values?

Consider the following normal distribution with the classic bell shaped curve. It is very important you understand that this is a histogram like the one you just drew. The bins are very small, meaning the columns are very narrow, giving the impression of a smooth curve. Many datasets, such as IQ-scores and SAT scores, will have this distribution or shape.

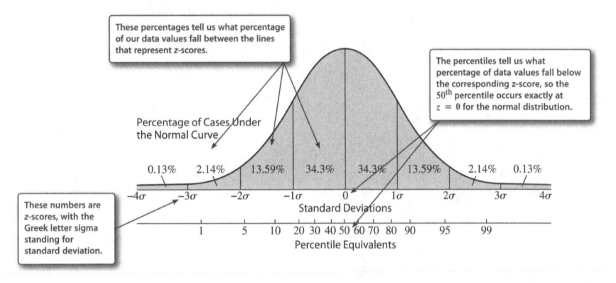

8. What percentage of people will have an IQ-score that is more than 1 standard deviation above the mean?

9. Notice that the normal distribution has less than 1% of data values outside of 3 standard deviations from the mean. What is the IQ-score of someone with a z-score of 3 (3 or higher is considered "genius")? What is the SAT score of someone with a z-score of 3?

Units, Conversions, Scales, and Rates

> **Definition:** In measuring a quantity, the **units** refer to the choice of measurement system. There are two distinct types of units:
> 1. **Units of observation**: These units pertain to a quality or category, such as nationality: American, Indian, Mexican, etc. A data value either belongs to a category or it does not.
> 2. **Units of measurement**: These units pertain to a quantity, such as length, using an agreed upon standard: feet, centimeters, miles etc. A data value for a given quantity can be measured using any associated unit of measurement, each of which can be converted to the other units of measurement.

Definition: A **conversion** is a ratio used to compare two different systems of measurement. A conversion indicates a change in units of measurement, which are typically proportional to one another. The constant of proportionality called the *conversion factor*.

1. 1 **km** = 0.6214 **miles** and 1 **mile** = 5,280 **feet**
   Convert 149.6 km to feet.

2. Write down the equation for this proportional relationship between km and ft.

3. Covert 149.6 million km to million feet.

Definition: There are two types of **scales** we encounter:

1. A *measurement scale* is a system of ordered marks, such as rulers and thermometers, used as a reference standard. These can be *ratio scales* or *interval scales*, depending on whether the units are *ratio units* or *interval units*.

2. A *model scale* is a ratio used to determine the size relationship between a model and that which it represents. A model scale indicates a change in magnitude. A model scale is typically given with the first quantity scaled to 1, and always between proportional quantities, with the constant of proportionality called the *scaling factor*.

4. **Planetary Scale:** You create a scale model of our solar system with the distance of the Earth to the Sun scaled to 1.

Planet	Distance to Sun (10^6 km)	Distance in Model
Mercury	57.9	
Venus	108.2	
Earth	149.6	1

Compute the distances to the sun of Mercury and Venus in the scaled model.

5. What are the units in the scaled model?

Definition: A **rate** is a ratio between quantities with different units, with the second quantity scaled to a "meaningful standard" and read using the word "per." Rates have *compound units* like meters per second (m/s) or miles per gallon (mi/gal).

In 2011 nearly 14 million monthly prescriptions were written for American adults 20-39 for ADHD. If there were 82,364 thousand adults aged 20-39 compute the following:

6. The rate of prescriptions per 100 adults 20-39.

7. The rate of prescriptions per 1,000 adults 20-39.

8. The rate of prescriptions per 100,000 adults 20-39.

9. Which rate do you think is "best"?

Definition: The **concentration** of a solution is the ratio of the amount of one substance mixed with the amount of another substance. This is typically a weight-to-volume ratio but can also be weight-to-weight. The **dosage** of a given medication is the ratio of amount of medicine to the weight of a patient.

Common Conversions

Weight			Volume		
1 gram (g)	=	0.001 kilograms (kg)	1 US fluid ounce	≅	29.57353 milliliters (ml)
1 gram (g)	≅	0.035273962 ounces	1 US cup	=	16 US tablespoons
1 ounce	=	28.34952312 grams (g)	1 US gallon	=	128 US fluid ounces
1 pound (lb)	=	0.45359237 kilograms (kg)	1 liter (l)	≅	33.8140227 US fluid ounces
1 kilogram (kg)	≅	35.273962 ounces	1 milliliter (ml)	=	1 cubic centimeter (cc)
1 kilogram (kg)	≅	2.20462262 pounds (lb)	1 US gallon	=	3.7854 liters

Converting Compound Units

10. Convert the density of water, 1 gram per ml, to pounds per gallon.

11. A doctor orders 2.3 g of Rocephin to be taken by a 15 lb. 6 oz. infant every 8 hours. The recommended dosage is 75-100 mg/kg per day. Is the ordered amount within the recommended range?

Rates, Canceling Units, and a Clever Equation

1. You have a solution mixed at a concentration of 5 g/ml and need 100 gal mixed at 4 g/ml. What do you do?

$$C_1 \cdot V_1 = C_2 \cdot V_2$$

2. Your roommate walks at 3 mph and it takes her 8 minutes to walk to class. You have the same class but walk faster at 5 mph. How long will it take you to get to class?

$$R_1 * T_1 = R_2 * T_2$$

3. You have a fixed budget for flooring in your new home. One type of flooring, at $4 per square foot, covers 89,280 square inches using your budget. The more expensive flooring at $6 per square foot will cover how much flooring on your budget?

$$R_1 * A_1 = R_2 * A_2$$

4. Jeannine eats ¼ cup oatmeal per day and a container lasts her 36 days. Her husband starts eating ½ cup per day. How long will the container now last with both eating oatmeal each day?

$$R_1 * T_1 = R_2 * T_2$$

5. A *mole* is a strange unit of measurement giving the amount of a substance that contains 6.022×10^{23} "elementary entities" (atoms, molecules, etc.). The *molarity* is a concentration measured in moles per liter. You have a solution with a molarity of 10^{-3} moles/L and need 100 ml at 10^{-4} moles/L. What do you do?

$$M_1 * V_1 = M_2 * V_2$$

6. You wish to count the number of bacteria in 1 ml of a solution but there are too many to count! So you mix the 1 ml with 9 ml of water to dilute it, and then draw 1 ml of the new mixture. Still too many, so you now mix with 9 ml of water and draw 1 ml of this new solution and now count 75 nasty little bacteria swimming around. How many bacteria are in the original solution?

$$C_1 \cdot V_1 = C_2 \cdot V_2$$

Z-scores, Standard Error, and the Normal Curve

The histograms we created earlier are discrete versions of idealized continuous distributions, such as the following normal distribution shown:

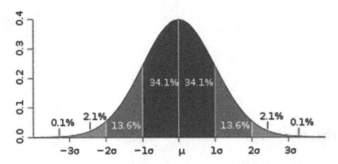

1. Recall that this curve is a **histogram**, just with really skinny columns (i.e. the bins are very small). What does the height of each column represent?

2. In the normal distribution what percentage of data values fall below (to the left of) the mean, μ?

   Below $+1\,\sigma$ (= +1 standard deviation)?

   Below $+2\,\sigma$?

   Below $+3\,\sigma$?

3. Here are some fun questions. Make an educated guess ☺:

    a. 95% of US women fall between what two heights?

    b. 95% of US men fall between what two heights?

4. On the given normal curve, what percentage of data values lie within two standard deviations of the mean (from -2σ to $+2\sigma$)?

Now consider the distribution of heights for men and women. These values were computed from a *random, representative sample* of men and women.

Table 3.9 Height in Inches for U.S. Females 20 Years of Age and Older by Race and Ethnicity and Age, Mean, Standard Error of the Mean, and Selected Percentiles, 2003-2006

Race and Ethnicity and Age	Number Examined	Mean	Standard Error	5th	10th	15th	25th	50th	75th	85th	90th	95th
All race and ethnicity groups								Inches				
20 years and over	4,857	63.8	0.06	59.3	60.3	61.0	62.1	63.8	65.6	66.6	67.2	68.2
20–29 years	1,061	64.3	0.12	59.9	60.9	61.6	62.5	64.2	66.1	66.9	67.5	68.0
30–39 years	842	64.3	0.13	60.0	60.8	61.5	62.5	64.2	66.0	67.1	67.7	68.6
40–49 years	784	64.2	0.12	59.9	60.6	61.4	62.4	64.2	66.0	66.9	67.7	68.5
50–59 years	604	63.9	0.13	59.3	60.4	61.2	62.2	63.8	65.7	66.4	67.1	67.9
60–69 years	691	63.7	0.13	59.8	60.5	61.1	62.1	63.7	65.3	66.1	66.9	67.5
70–79 years	463	62.7	0.13	58.6	69.4	60.1	61.0	62.6	64.4	65.2	65.9	66.7
80 years and over	412	61.4	0.15	57.5	58.3	58.8	59.7	61.3	62.9	63.9	64.7	65.4

Data from: National Health Statistics Report No. 10, October 2008, Anthropometric Reference Data, U.S. 2003–06

Table 4.21 Height in Inches for U.S. Males 20 Years of Age and Older by Race and Ethnicity and Age, Mean, Standard Error of the Mean, and Selected Percentiles, 2003-2006

Race and Ethnicity and Age	Number Examined	Mean	Standard Error	5th	10th	15th	25th	50th	75th	85th	90th	95th
All race and ethnicity groups								Inches				
20 years and over	4,482	69.4	0.07	64.4	65.6	66.3	67.4	69.4	71.5	72.6	73.2	74.3
20–29 years	808	69.4	0.13	64.7	65.8	66.6	67.8	70.0	72.0	73.0	73.5	74.8
30–39 years	742	69.4	0.13	64.1	65.3	66.1	67.5	69.5	71.5	72.7	73.4	74.7
40–49 years	769	69.7	0.11	65.2	66.2	66.8	67.9	69.7	71.6	72.7	73.3	74.0
50–59 years	591	69.5	0.15	65.0	65.8	66.5	67.5	69.5	71.5	72.7	73.4	74.4
60–69 years	668	69.0	0.11	64.2	65.4	66.1	67.1	69.0	71.1	71.9	72.7	73.7
70–79 years	555	68.4	0.16	63.8	64.6	65.5	66.4	68.5	70.3	71.0	72.0	73.1
80 years and over	349	67.2	0.14	62.7	63.6	64.3	65.5	67.2	68.9	70.0	70.5	71.3

Data from: National Health Statistics Report No. 10, October 2008, Anthropometric Reference Data, U.S. 2003–06

Definition: The **standard error of the mean** is the standard deviation of the distribution of sample means. The standard error can be estimated using a standard deviation from one of the samples: $SE \cong SD / \sqrt{N}$, where N is the number of values in the sample.

Caution! The standard error in the given tables was computed using a more sophisticated formula, so you cannot compute SD from SE for these tables.

The Standard Error for women 20 years and over is given as 0.06 inches and can be interpreted using the fact that 68% of data in a normal distribution will lie within 1 standard deviation of the mean as follows:

- There is a 68% probability that the average height of women 20 years and over is 63.8 inches $\pm$ 0.06 inches.
- There is a 68% chance that 63.8 inches is within 0.06 inches of the true average height (or population average) of women 20 years and over.
- There is a 68% likelihood that the average height of women 20 years and over lies in the interval from 63.74 inches to 63.86 inches.

5. Estimate the following probabilities using the percentages and SE in the table:

    a. The probability that a random 20-29 year old woman is over 66.9 inches tall.

    b. The probability that a random sample of 20-29 year old women has a mean height over 66.9 inches.

6. Interpret the **standard error** for males 20 years and over in a sentence using 95% probability.

7. Recall that 85% of the data fall below $+1\sigma$ so we can determine the **standard deviations** for men and women using the 85th percentile. Determine the SD by subtracting the mean height from the 85th percentile for both men and women 20 years and over.

8. Determine the z-score for your height.

9. Use the standard deviations to determine the heights between which 95% of men and women fall. How close were your guesses?

10. Who is taller: a 6 foot tall woman or a 6 foot 6 inch tall man?

Normal Distributions

a.k.a. *The Bell Shaped Curve, The Normal Curve, The Gaussian, The Great Intellectual Fraud (GIF)*

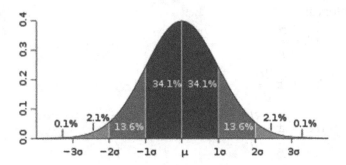

There are 2 key characteristics of the normal distribution and its associated *bell shaped curve*:

1) **Area under the curve is 1:** Note that each column in the histogram (frequency distribution) has an area equal to a probability, and the sum of the probabilities equals 1!

2) **Perfect symmetry about the mean:** Note that this does not imply the standard deviation is the same for every normal distribution; the bell shaped curve can have different centers and spreads.

Most amazing and useful fact: For any **normal distribution**, regardless of the center (mean) or spread (standard deviation) determine the appropriate percentages:

- ____% of the data values lie within **1 standard deviation** from the mean.

- ____% of the data values lie within **2 standard deviations** from the mean.

- ____% of the data values lie within **3 standard deviations** from the mean.

Important results for any, including **non-normal**, distributions:

- **Chebyschev's Theorem:** For any data set at least 75% of the data values are within 2 standard deviations of the mean and at least 89% are within 3 SD's.

- **Range rule of thumb:** For any data set the standard deviation is approximately one-fourth of the range (largest data value minus the smallest).

- **Central Limit Theorem:** For any data set the distribution of means from many samples (more than 30) of the same size is approximately normal. The mean of this distribution is the true mean of the population, μ, and the standard deviation is the population SD divided by the square root of the sample size, SE = $\dfrac{\sigma}{\sqrt{n}}$, which is the **standard error of the mean** introduced in Chapter 3.

95% rule and the Central Limit Theorem: Given a population with mean, μ, standard deviation, σ, and a representative random sample of size, n, with mean $\bar{X}$ and standard deviation, SD, the following statements are equivalent:

- 95% of the sample means are in the interval

$$\mu \pm \frac{2\sigma}{\sqrt{n}} \; .$$

- There is a 95% chance that $\qquad \bar{X} \in \mu \pm \dfrac{2\sigma}{\sqrt{n}} \; .$

- There is a 95% chance that $\qquad \mu \in \bar{X} \pm \dfrac{2SD}{\sqrt{n}} \; .$

Note the switch between population and sample standard deviations.

The Black Swan by Nassim Taleb
Why does he call the Normal Distribution the Great Intellectual Fraud?
Because socio-economic data is usually not normal but economists' models all use the normal distribution!

Example: US Household Income 2005

$\mu = \$84,800$ and $\sigma = \$386,000$ Note: median value = \$58,500

1. What does the mean being greater than the median tell you about the data set?

2. The top 0.01% of households averaged \$35,473,200. Compute the corresponding z-score.

3. U.S. Women 20-29 years old have an average height of 64.1 inches with a standard deviation of 2.9 inches. How tall would a woman be if she had the same z-score computed in **#2**?

4. What is a "black swan" a metaphor for in economic terms?

Labor Picture December 2014: Percentages Worksheet

> **Definition:** A **percentage** is a ratio where the second quantity has been scaled to 100, x : 100. The first quantity is then said to be x percent (%) of the second. Percentages are sometimes referred to as "rates" because they are "*per 100.*"

Data from: Bureau of Labor Statistics, U.S. Department of Labor, Employment Situation News Release, http://www.bls.gov/news.release/pdf/empsit.pdf

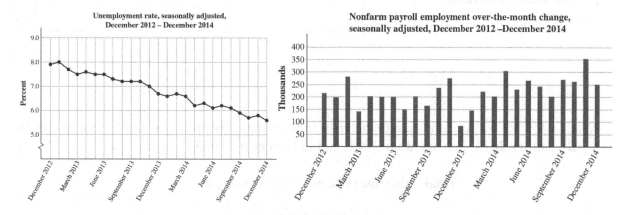

1. The U.S. civilian non-institutional population was 249 million, how big was the *labor force* in December **2014** and how many people were employed?

Share of Civilian Non-Institutional Population			
	DEC.	1-Month Change	1-Year Change
Employed	59.2%	Unch.	+0.6 pts.
Labor Force (16 and over Workers and Unemployed)	62.7	-0.2	-0.1

2. What percentage of the U.S. civilian non-institutional population was *employed* in December of **2013**?

$$x\% \xrightarrow{\text{+0.6 pts}} 59.2\%$$

2013 (Original)   2014 (New)

3. Under *Hidden Unemployment*, how many people in December **2014** were *working part-time but looking for full-time work?*

"Hidden" Unemployment			
In millions	DEC.	1-Month Change	1-Year Change
Working part-time , but want full-time work	6.8	-0.9%	-12.6%
People who currently want a job (includes discouraged workers)	3.0	+6.9	-10.3

4. Use the following formulas to determine how many people *currently want a job* in **December 2014**, in **November 2014**, and in **December 2013**.

Total (Absolute) Change = New – Original.

$$\text{Percent (Relative) Change} = \frac{\textbf{Total Change}}{\text{Original}}$$

Same equation!

New = Original $\pm$ *x* % $\cdot$ Original

$$\text{New} = \text{Original} \cdot \left(1 \pm \frac{x}{100}\right)$$

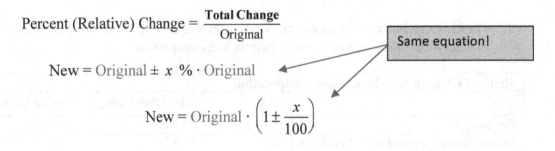

$x \xrightarrow{\hspace{4cm}} 3.0$

11/14 (Original) +6.9%

12/14 (New)

$$x \xrightarrow{\hspace{1cm} -10.3\% \hspace{1cm}} 3.0$$

12/13 (Original)   12/14 (New)

5. What was the average (mean) duration of unemployment in weeks in December 2014? What was this in **November 2014**? **December 2013**?

Duration of Unemployment			
In weeks	DEC.	1-Month Change	1-Year Change
Average	32.8	-0.6%	-10.9%
Median	12.6	-1.6	-25.9

6. Why is the average duration of unemployment greater than the median duration?

7. In the *Demographics* section which is true and why:
a. 4.8% of Whites are unemployed; b. 4.8% of unemployed are White.

Unemployment Demographics			
	DEC.	1-Month Change	1-Year Change
White	4.8%	-0.1 pts.	-1.2 pts.
Black	10.4	-0.6	-1.4
Hispanic	6.5	-0.1	-1.9
Asian	4.2	-0.6	+0.1
Teenagers (16-19)	16.8	-0.7	-3.6

8. Which *Demographic* group (excluding Asians) experienced the greatest 1 year percentage decrease in unemployment? (Note that the chart gives you total change.)

9. In the *Unemployment by Education Level* section can you add the percentages for High School and less than High School and say: "13.9% of the unemployed have at most a high school education"?

Unemployment by Education Level			
	DEC.	1-Month Change	1-Year Change
Less than high school	8.6%	+0.1	-1.3 pts.
High school	5.3	-0.3	-1.7
Some College	4.9	Unch.	-1.2
Bachelor's or higher	2.9	-0.3	-0.5

10. What is the growth **factor** for the 1 year change in average weekly earnings?

Average Weekly Earnings			
	DEC.	1-Month Change	1-Year Change
Rank and file workers	$850.12	-0.2%	+2.5%

Proportionality and Linear Functions

This worksheet connects the concepts of proportionality and constant rate to linear functions. Please read through the first example that has been done for you, and then complete the following three examples.

1. **Monthly cell phone bill**: Cost is proportional to minutes you speak at a rate of 5 cents per minute.

   a. Write an equation for this proportional relationship.

   $$\text{Cost} = 0.05 \cdot \text{Minutes}$$

   b. Now assume there is an additional fixed cost of $20 each month. Write down the new equation.

   $$\text{Cost}_2 = 0.05 \cdot \text{Minutes} + 20$$

   c. Fill in table of values for both functions:

MINUTES	COST	COST + $20
0	$ -	$ 20.00
100	$ 5.00	$ 25.00
200	$ 10.00	$ 30.00
300	$ 15.00	$ 35.00
400	$ 20.00	$ 40.00
500	$ 25.00	$ 45.00

   d. Plot points and sketch the graph of each function.

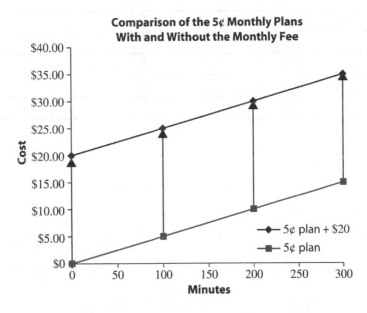

e. How does the constant rate of 5 cents per minute show up in the graph?

The rate can be thought of as $5 per 100 minutes, so to get from one point to the next we go up $5 (RISE) and right 100 minutes (RUN). This rate is referred to as the SLOPE of the line because it quantifies the steepness of the line, and is often remembered as the ratio of RISE over RUN.

Adding in a $20 monthly fee raises the graph of the line up 20 units. Note that it does not affect the slope, or the steepness, of the line

2. **Weight of water:** The weight of water is proportional to the volume at a rate of 8 pounds per gallon.

   a. Write an equation for this proportional relationship.

   b. Now assume the bucket you are using to carry the water weighs 10 pounds. Write down the new equation.

   c. Fill in table of values for both functions:

V (gal)	W (lbs)	W(lbs) + 10 lbs.
0		
0.5		
1.0		
1.5		
2.0		
2.5		

d. Plot points/ sketch graph of both functions:

e. How does the constant rate of 8 lbs. per gallon show up in the graph?

3. **Buying tickets:** The cost is proportional to the number of tickets you buy at the fair at a rate of $2 per ticket.

    a. Write an equation for this proportional relationship.

    b. Now assume there is a fixed cost of $15 to get into the fair. Write down the new equation.

c. Fill in table of values for both functions:

T	c (\$)	C (\$)
0		
5		
10		
15		
20		
25		

d. Plot points and sketch the graph of each function.

e. How does the constant rate of \$2 per ticket show up in the graph?

4. **Driving distance:** The distance you drive is proportional to the time driving at a constant speed of 40 miles per hour.

    a. Write an equation for this proportional relationship.

    b. Now assume the distance driven includes the 100 miles you already drove. Write down the new equation.

    c. Fill in table of values for both functions:

T (hours)	d (miles)	D (miles)
0		
1		
2		
3		
4		
5		

    d. Plot points and sketch graph of each function.

e. How does the constant rate of 40 miles per hour show up in the graph?

Linear Functions

Definition: Let (x_1, y_1) and (x_2, y_2) be points in the coordinate plane such that $x_1 \neq x_2$. The **slope** of the straight line that passes through these two points is given by:

$$m = \frac{y_2 - y_1}{x_2 - x_1} = \frac{\Delta y}{\Delta x} = \frac{\text{Rise}}{\text{Run}} = \frac{\text{Total Change in OUTPUT}}{\text{Total Change in INPUT}}.$$

The **slope** is a **ratio** of the total change in the OUTPUT to the total change in the INPUT.

The following table gives the amount of money spent on Social Security in this country:

Year	Years Since 1975	Social Security Outlays (millions $)
1975		64,658
1980		118,547
1985		188,623
1990		248,623
1995		335,846
2000		409,423
2005		523,305
2012		

1. Take the first and last data points: (1975, $64,658) and (2005, $523,305) and compute the slope. Interpret as a *rate*: something per something.

2. Draw the line between the two points on the following graph:

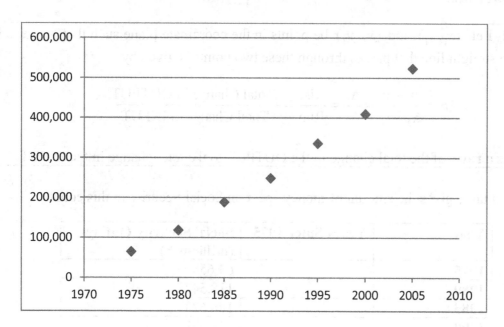

3. Compute the equation of the line: $y = mx + b$, by plugging in an (x,y) point and the slope, m, and solving for b.

4. Where does your line look like it will hit the vertical axis on the graph? What year does this vertical axis correspond to? How does this relate to the value of b in your equation in **#3**?

5. Now fill in the Years Since 1975 column in the table and put the corresponding numbers on the x-axis of the graph. Draw a vertical line through $x=0$ on the graph. This is your y-axis.

Year	Years Since 1975	Social Security Outlays (millions $)
1975		64,658
1980		118,547
1985		188,623
1990		248,623
1995		335,846
2000		409,423
2005		523,305
2012		

6. Compute the equation of the line through the points (0, $64,658) and (30, $523,305). How does this differ from the equation in **#3**?

7. Now consider the *trendline* (*best fit* line or *least-squares* line) in the following graph. How does this line differ from the one you drew?

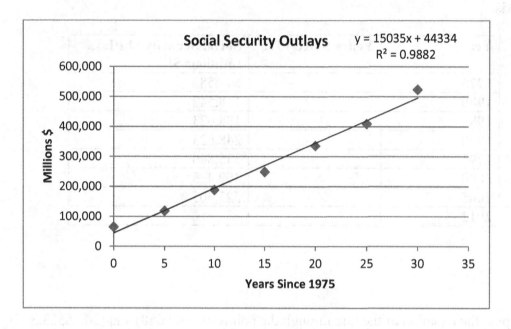

8. Use both your line (x = years since 1975) and the best fit line to predict Social Security Outlays in 2012. Which is closer to the actual amount of $808.04 billion?

Divorce Rate

The following spreadsheet compares various divorce and marriage rates.

	A	B	C	D	E	F	G	H	I	J
1										
2		US	Number of	Number of	Total Number	Divorces	Marriages	Divorces per	Divorces per	
3		Population	Divorces	Marriages	Marriages in the	in that year per	in that year per	100	1,000	
4			Occurring	Occurring	Population	1,000	1,000 Marriages	Marriages		
5	Year		in that year	in that year		Population	Population	in that year	in the population	
6	1900	75,994,575	55,751	685,284	13,885,000	0.7	9.0	8.1	4.0	
7	1910	91,972,266	83,000	948,000	17,843,000	0.9	10.3	8.8	4.7	
8	1920	105,710,620	171,000	1,274,000	21,333,000	1.6	12.1	13.4	8.0	
9	1930	122,775,046	183,000	1,102,000	26,245,000	1.5	9.0	16.6	7.0	
10	1940	131,669,275	269,000	1,566,000	30,150,000	2.0	11.9	17.2	8.9	
11	1950	150,697,361	385,000	1,667,000	37,450,000	2.6	11.1	23.1	10.3	
12	1960	178,464,236	393,000	1,523,000	42,200,000	2.2	8.5	25.8	9.3	
13	1970	203,211,926	708,000	2,159,000	47,600,000	3.5	10.6	32.8	14.9	
14	1980	222,300,000	1,189,000	2,390,000	52,300,000	5.3	10.8	49.7	22.7	
15	1990	250,132,000	1,182,000	2,443,000	56,300,000	4.7	9.8	48.4	21.0	
16	2000	282,339,000	1,157,589	2,329,000	60,050,000	4.1	8.2	49.7	19.3	
17	2004	296,410,000	1,126,358	2,311,998	61,410,000	3.8	7.8	48.7	18.3	
18										
19										

Data from: U.S. Dept. of Health and Human Services, National Center for Health Statistics. Web: www.cdc.gov/nchs/

1. For the year 2004 use each of the four rates in a sentence.

2. Which of these rates do you feel is the best way to report divorce rates? Explain.

3. Which rates seems to support the idea that half of all marriages end in divorce?

4. Does the conclusion that roughly half of all marriages end in divorce logically follow from these rates?

5. Consider the following $x - y$ scatterplot. In what year does the linear trend radically change?

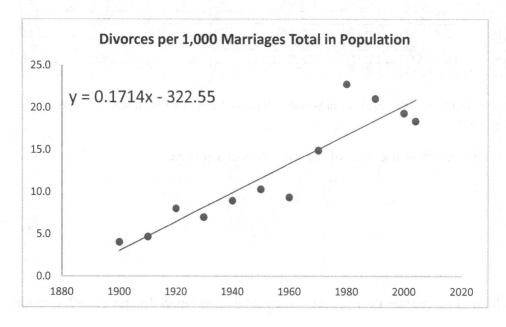

Divorces per 1,000 Marriages Total in Population

$y = 0.1714x - 322.55$

6. What could explain the divorce rate starting to fall after 1980?

7. If you knew that the number of new marriages has been declining since 1980, as more and more young couples choose to co-habitate, does that change your answer to #6?

8. Interpret the slope of the line as a rate.

9. Find the equation of the line going between the points (0, 22.7) and (24, 18.3) using x = years since 1980. Use your line to predict the divorce rate per 1,000 marriages total in the year 2020.

Exponential Growth

Carbon-14 dating

Carbon-14 is an element that naturally builds up in pants and animals during life, and then will decay exponentially after death at a rate of 11.4% per millennia.

1. Using the equation, Factor = 1 ± Rate, determine the decay factor for Carbon-14.

2. Assuming you have 100 grams of Carbon-14 to start fill in the following table:

MILLENIA (1,000's years)	CARBON-14 REMAINING (g)
0	100
1	
2	
3	

3. Write down an equation for the amount of Carbon-14 remaining, C, after m millennia.

4. The following graph shows how Excel represents this equation. What is the continuous decay rate for Carbon-14?

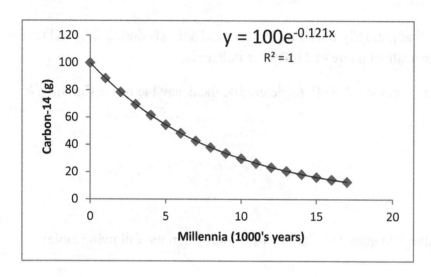

$$y = 100e^{-0.121x}$$
$$R^2 = 1$$

5. Use the equation from Excel and your e^x button on your calculator (or **EXP** function in Excel) to compute the amount of Carbon-14 left after 8,250 years.

Geometric Mean and Average Percentage Change

Average percentage change requires the use of the *geometric mean*. Imagine a town with a population of 900 that grows by 30% in 1 year and then 72% in the next. What is their average percentage change over the 2 years?

Caution! The average percentage change is NOT $\dfrac{30\% + 72\%}{2} = 51\%$. This uses the *arithmetic mean* and does not work in this situation.

The CORRECT way to solve this is using growth factors. The town starts at 900 people and then the population is multiplied by the growth factor of 1.30 and then multiplied by 1.72:

$$900 \times 1.30 \times 1.72 = 2{,}012 = 900 \times (1 + r) \times (1 + r)$$

We now have two ways to proceed:

Solution Procedure #1: Use just the growth factors to solve for the average growth factor and then subtract 1 to find the average growth rate, r.

$$(1 + r)^2 = 1.30 \times 1.72$$

$$1 + r = \sqrt{1.30 \times 1.72} = 1.495$$

$$r = 49.5\%$$

$$1 + r = \sqrt[n]{(1 + r_1) \cdot (1 + r_2) \cdots (1 + r_n)}$$

Solution Procedure #2: Use the *Original* and *New* values to solve for the average growth factor and then subtract 1 to find the average growth rate, r.

$$(1 + r)^2 = \frac{2{,}012}{900}$$

$$1 + r = \sqrt{\frac{2{,}012}{900}} = 1.495$$

$$r = 49.5\%$$

$$1 + r = \sqrt[n]{\frac{\text{New}}{\text{Original}}}$$

Use the appropriate technique to answer the following:

6. A mutual fund increases 8% in 1 year, loses 5.2% in the next, and then is up 12.3% the following year. What is the average percentage change?

7. A town grows from 12,456 people to 32,980 in 5 years.
   a. What is the average total change per year?

   b. What is the average percentage change per year?

Exponential Puzzlers

1. You win a radio show contest and have 1 minute to choose between receiving $10,000 or you can get the sum of the following payouts for 30 days:

DAY 1	1 CENT
Day 2	2 cents
Day 3	4 cents
Day 4	8 cents
… continue doubling	
Day 30	
SUM:	???

a) How much money do you get on Day 5?

b) How much money do you get on Day 30?

c) How much do you get in total? Hint: Use a spreadsheet.

2. A mutual fund averages 7.2%, -4.3%, and -9.1% over 3 years. What was the average rate of return over the 3 years?

3. You mix equal weights (30 kg) of lead at 11,340 kg/m^3 and gold at 19,300 kg/m^3. What is the density of the resulting alloy?

4. Apple Inc. had a net profit of $57 million in 2003 and a net profit of $14,013 million in 2010. Wow! What was the average change of Apple's net profit over this 7 year period?
   a. In $ per year?

   b. As % change per year?

The Parable of the Bacteria in a Bottle[1]: Once upon a time, at precisely 11:00 pm, a single bacterium was placed into a nutrient-filled bottle in a laboratory. The bacterium immediately began gobbling up nutrients, and after just one minute — making the time 11:01 — it had grown so much that it divided into two bacteria. These two ate until, one minute later, they each divided into two bacteria, so that there were a total of four bacteria in the bottle at 11:02. The four bacteria grew and divided into a total of eight bacteria at 11:03, sixteen bacteria at 11:04, and so on. All seemed fine, and the bacteria kept on eating happily and doubling their number every minute, until the "midnight catastrophe." The catastrophe was this: At the stroke of 12:00 midnight, the bottle became completely full of bacteria, with no nutrients remaining — which meant that every single one of the bacteria was suddenly doomed to death.

We now turn to our questions, as we seek to draw lessons from the tragic demise of the bacterial colony.

Question 1: The catastrophe occurred because the bottle became completely full at 12:00 midnight. When was the bottle *half*-full?

Question 2: You are a mathematically sophisticated bacterium, and at 11:56 you recognize the impending disaster. You immediately jump on your soapbox and warn that unless your fellow bacteria slow their growth dramatically, the end is just 4 minutes away. Will anyone believe you?

---

[1] Adapted from Jeffrey Bennett's excellent book, *Math for Life*.

Question 3: It's 11:59 and your fellow bacteria are finally taking your warnings seriously. Hoping to avert their impending doom, they quickly start a space program, sending little bacterial spaceships out into the lab in search of new bottles. To their relief, they discover that the lab has four more bottles that are filled with nutrients but have no one living in them. They immediately commence a mass migration through which they successfully redistribute the population evenly among the four bottles just in time (at midnight all 4 bottles are one-quarter full) to prevent the midnight catastrophe. How much more time do the additional bottles buy for their civilization?

Question 4: Because the four extra bottles bought so little time, the bacteria keep searching out more and more bottles. Is there any hope that additional discoveries will allow the colony to continue its rapid growth? Assume each bacterium occupies a volume of 10^{-21} cubic meters and the surface area of the Earth is 510 million square km. At 1:00 AM how deep are the bacteria if they are evenly spread over the surface of the earth?

Logarithms

The total amount of contaminant in a pond decreases by 25% each day.

1. Assuming you started with 100 ppm in the water, how long would you estimate until it is half gone?

2. Fill in the following table:

Days	0	1	2	3	...	t
Amount (ppm)						

3. Write down the equation for solving when the amount will be 50, exactly one-half of the original.

4. Compute without calculator the following:
   a. $\log_2 8 =$ _____

   b. $\log_3 9 =$ _____

   c. $\log_{10} 10,000 =$ _____

   d. $\log 10^{-1} =$ _____

   e. $\ln e^2 =$ _____

Crucial **PROPERTY** of Logarithms used to solve for Half-life or Doubling Time:

$$\log_b a^k = k \cdot \log_b a$$

5. Solve for the "half-life" using logarithms.

6. Solve for daily continuous decay rate using logarithms.

7. Fun Balderdash Question ☺

A **logarithm** is …
_____

8. Fill in the following table:

ANNUAL G/D RATE	DOUBLING/ HALF TIME IN YEARS	G/D FACTOR PER YEAR	CONTINUO US RATE PER YEAR	G/D FACTOR PER DECADE	G/D RATE PER DECADE
-25%		0.75		0.0563	
		1.043	4.21%		

AP...R AP...Y Oh My!

1. Write down the exponential equation for the amount of money, **P**, after **m** months at 8% compounded monthly starting with $1,000.

2. Note the relationship between Months and Years by filling in the following table:

Months (**m**)	12	48	6	3	**m = 12•t**
Years (**t**)					

3. Rewrite your equation from **#1** using years, **t**, as the input.

4. Check that $P(12)$, $m = 12$, from question **#1** equals $P(1)$, $t = 1$, from question **#2**.

$$P = P_0 \cdot \left(1 + \frac{APR}{n}\right)^{nt} \ \rightarrow\rightarrow\rightarrow\rightarrow \ P = P_0 \cdot \left[\left(1 + \frac{APR}{n}\right)^{n}\right]^{t} \ \rightarrow\rightarrow\rightarrow\rightarrow \ P = P_0 \cdot \left(1 + APY\right)^{t}$$

Definition: The **yield** or Annual Percentage Yield (**APY**) of an investment is the effective interest rate required to create the total amount of interest generated by an investment or loan over the course of 1 year. For an investment with no additional deposits or payments we have:

$$1 + APY = \left(1 + \frac{APR}{n}\right)^{n}$$

where n = the number of periods in 1 year.

	A	B	C	D	E	F	G
1	APR	8%	8%	8%	8%	8%	8%
2	NPer in 1 Year	2	4	12	26	52	365
3	Periodic Rate	4.000%	2.000%	0.667%	0.308%	0.154%	0.022%
4		Periodic Balance					
5	Periods	Bi-annually	Quarterly				
6	0	$1,000.00	$1,000.00	$1,000.00	$1,000.00	$1,000.00	$1,000.00
7	1	$1,040.00	$1,020.00	$1,006.67	$1,003.08	$1,001.54	$1,000.22
8	2	$1,081.60	$1,040.40	$1,013.38	$1,006.16	$1,003.08	$1,000.44
9	3	$1,124.86	$1,061.21	$1,020.13	$1,009.26	$1,004.62	$1,000.66
10	4	$1,169.86	$1,082.43	$1,026.93	$1,012.36	$1,006.17	$1,000.88
11	Years						
12	0	$1,000.00	$1,000.00	$1,000.00	$1,000.00	$1,000.00	$1,000.00
13	1	$1,081.60	$1,082.43	$1,083.00	$1,083.15	$1,083.22	$1,083.28
14	2	$1,169.86	$1,171.66	$1,172.89	$1,173.22	$1,173.37	$1,173.49

5. What formula is in cell **B3** that can be filled across?

6. Fill in the names for the **Periods** in cells **D5:G5**.

7. The formula in cell **B7** uses the basic idea of exponential growth with a growth rate given by the **Periodic Rate**. What formula is in cell **B7** that can be filled down and across?

Recall that you have two choices: a *closed formula* starting at $1,000 and using the number of periods as input; or a *recursive formula* that multiplies the previous by the growth factor (1 + rate).

8. Use your formula from question 7 to write down the formulas that will appear in **B8** and **C7** in order to check if you are right!

B8=

C7=

9. The formula in cell **B13** uses the equation $P = P_0 \cdot \left(1 + \dfrac{APR}{n}\right)^{nt}$ where t = years and n = the number of periods in 1 year. What formula is in cell **B13** that can be filled down and across?

10. Use your formula from question 9 to write down the formulas that will appear in B14 and C13 in order to check that you are right!

B14=

C13=

11. How much interest is charged after 1 year for each **Periodic Rate**?

12. What is the **APY** for each **Periodic Rate**?

Log Scales and Cumulative Frequency Distributions

Fill in the following table for a hypothetical data set:

GRADE BINS	COUNTS	RELATIVE FREQUENCY	CUMULATIVE
5...6.5	20	10.0%	10.0%
6.5...7.5	50	25.0%	35.0%
7.5...8	80		
8...9.5	40		
9.5...10	10		

1. For each column in the following histogram, write inside the column the data value associated to it.
2. Draw a curve connecting the tops of the shaded columns and another curve connecting the tops of the hashed columns. These curves show the "*distribution*" of grades.
3. For the *7.5 .. 8* bin, use the 40% and the 75% in a meaningful sentence.

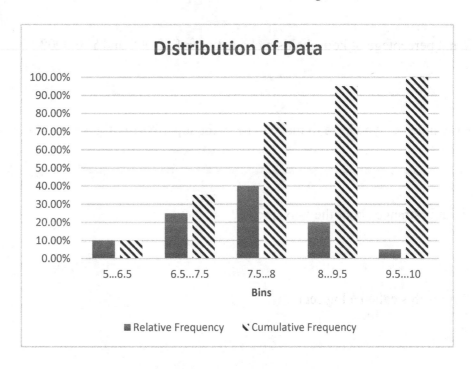

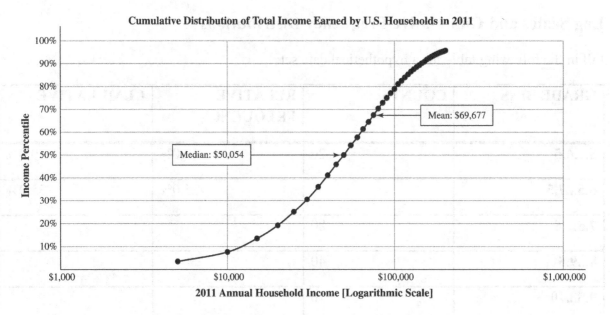

Cumulative Distribution of Total Income Earned by U.S. Households in 2011

Income Percentile (y-axis): 10% to 100%

Median: $50,054

Mean: $69,677

2011 Annual Household Income [Logarithmic Scale] (x-axis): $1,000 to $1,000,000

Data from: U.S. Census Bureau, Current Population Survey, 2012 Annual Social and Economic Supplement, Table HINC-01 (http://www.census.gov/hhes/www/cpstables/032012/hhinc/hinc01_000.htm)

4. This graph shows the household income distribution in the United States in 2011. What percentage of households make less than $100,000?

5. What percentage of households make between $50,000 and $70,000?

6. How much do you have to make to be in the top 10% of all household incomes?

7. What is strange about the x-axis scale?

8. Why is this called a log scale?

Correlation

"Correlation does not imply causation!"

– every stats teacher IN THE WORLD

The following table gives selected crude oil prices (per barrel) and gas prices (per gallon) with a scatter-plot of the first two columns:

	A	B	C
1	Date	Price per barrel	Price per gallon
2	1955	$26.06	$0.99
3	1960	$35.40	$1.19
4	1965	$37.58	$1.29
5	1970	$40.06	$1.49
6	1975	$47.71	$1.79
7	1980	$49.90	$1.89
8	1985	$54.36	$2.09
9	1990	$56.85	$2.19
10	2000	$61.68	$2.35
11	2005	$69.09	$2.85
12	2008	$145.16	$4.06
13	2009	$70.18	$2.65
14	2010	$87.65	$2.78
15			

1. Notice the two "**outlier**" points at the end that mess up the nice linear trend. Ignore those two points and sketch in a linear trendline.

2. What is your slope? Interpret using proper units. Compare your line to the "best-fitting" line in the scatterplot. In this chapter we will learn how this line is chosen!

Now consider the following graphs comparing oil prices to gas prices. The first one includes the two outlier points and the second one does not.

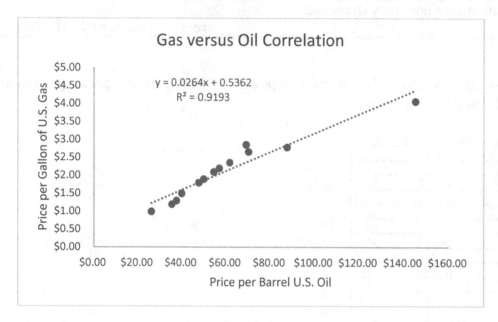

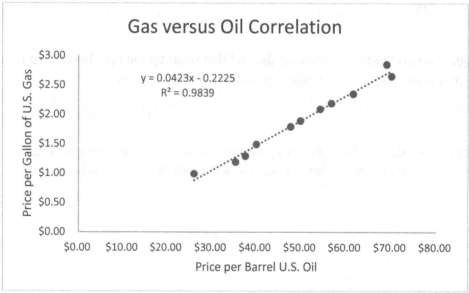

3. Interpret both slopes using proper units.

Notice the second data-set looks more linear (points are closer to the trend-line), and the "**R-squared**" value is indicating this. We can interpret this "**coefficient of linear determination**" in two ways:

- "The bivariate data, oil and gas prices, is 98.39% linear."
- "98.39% of the variability in gas price can be attributed to variability in oil price."

The **correlation coefficient**, R, is a measure of the strength and direction of the linear relationship for bivariate data and will be between -1 and 1, as indicated in **Table 8.2**.

The **coefficient of linear determination**, R^2, is a measure of linearity and will be between 0 and 1, with a value of 1 indicating perfect linearity. Technically, it is the proportion of variability of one variable in bivariate data that can be attributed to variability in the other variable.

R^2 can easily be interpreted as a percentage (which is why Excel reports R^2 and not R in scatterplots), using the following important sentence:

$R^2 \cdot 100\%$ *of the variability in the output can be attributed to variability in the input.*

Table 8.2 Interpretations of the Correlation Coefficient, R

How R Measures the Strength of the Relationship Between Two Variables				
Very Strong, Negative	Moderately Strong, Negative	Absolutely No Relationship	Moderately Strong, Positive	Very Strong, Positive
-1.0	-0.5	0	0.5	1.0

Now consider the relationship between strength of schedule (SOS) and the percentage of games won for women's Division III basketball.

4. Do you think this will be a positive relationship (harder the schedule more games won) or negative relationship? Explain!

5. What percentage of the variability in games won do you think is due to variability in the strength of schedule?

6. Check your guesses against actual data from the 2009-10 Division season:

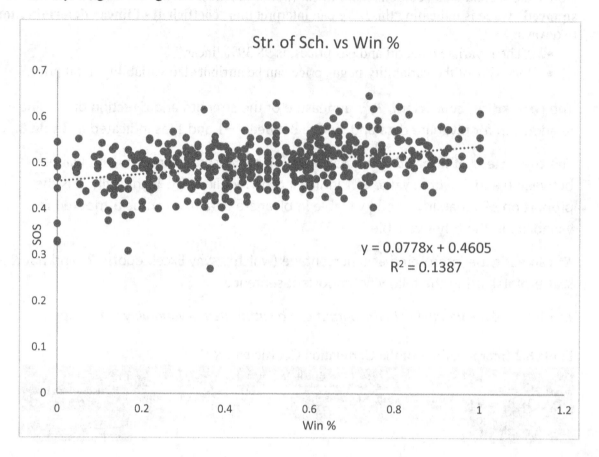

Str. of Sch. vs Win %

$y = 0.0778x + 0.4605$

$R^2 = 0.1387$

Line of Best Fit

The Fishy Data set consists of 42 length/weight measurements of trout from the Spokane River.

	A	B	C	D	E
1	Rainbow trout from the Spokane River, WA (n = 42)				
2	Source: WA State Dept. of Ecology report				
3	length (mm)	weight (g)		length (mm)	weight (g)
4	457	855		270	209
5	405	715		359	476
6	455	975		347	432
7	460	895		259	202
8	335	472		247	184
9	365	540		280	248
10	390	660		265	223
11	368	581		309	392
12	385	609		338	460
13	360	557		334	406
14	346	433		332	383
15	438	840		324	353
16	392	623		337	363
17	324	387		343	390
18	360	479		318	340
19	413	754		305	303
20	276	235		335	410
21	387	538		317	335
22	345	438		351	506
23	395	584		368	605
24	326	353		502	1300

Let X = length of trout (mm) and Y = weight of trout (g). The following statistics have been computed for you:

	MEAN $\overline{X}_{length}$ or $\overline{Y}_{weight}$	STANDARD DEVIATION SD_X or SD_Y
X = length (mm)	352.90	57.02
Y = weight (g)	501.02	230.51

R_{XY}, the correlation coefficient, is 0.969.

1. Compute the least squares line using the equation:

$$y = \bar{Y} + R \cdot \frac{SD_Y}{SD_X} \cdot \left(x - \bar{X} \right)$$

2. Interpret the slope as a rate of change.

3. Interpret the y-intercept.

4. Why does the y-intercept make no sense? Look at the following scatterplot. For what domain does this linear model seem valid?

5. You catch a whopper, measuring 484 mm in length! What is the z-score of this trout?

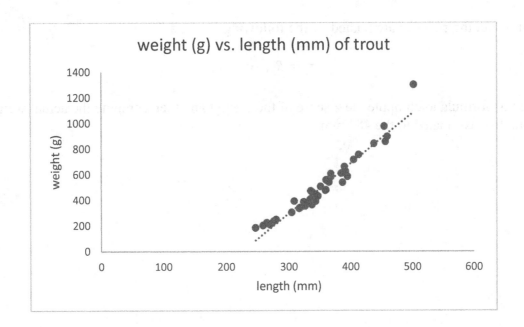

weight (g) vs. length (mm) of trout

6. On the scatterplot, mark the means, $\overline{X} = 352.9$ and $\overline{Y} = 501.02$, and the 484 mm on the respective axes.

7. What would you estimate the weight of this BIG fish to be? Place a mark on the y-axis for this weight.

8. What is the z-score of this weight? How is it related to the z-score of the length?

9. It turns out the z-scores are related by the following formula:

$$z_Y = R_{XY} \cdot z_X$$

Use this formula to compute the z-score of the weight and then compute the actual weight of the fish associated to the 484 mm.

RNPPF
Rover Needs Poopy Paper Fast!

Use the **RATE, NPER, PMT, PV** and **FV** functions to answer the following. Remember the following important subtleties:
- RATE = periodic rate; you must divide the APR by the number of periods in one year.
- Anytime you send money to the bank it must be entered as a negative.
- It is assumed all PMT's are made at the END of each period, while PV is the amount you have to START. (You may use the TYPE argument to change PMT's to the beginning of the period.)

Write down the function as you would type into Excel:

1. What is your monthly payment on a $23,000 car loan at 8.75% for 5 years?

2. What APR will you need to guarantee yourself an annual payment of $60,000 for 25 years if you have $1,200,000 in the bank and you anticipate having nothing left at the end of 25 years?

3. How much money will you end up with if you make $650 monthly payments to your retirement account which earns 11.3% for 36 years?

4. Compute the following without typing them into Excel:
   a. = **FV** (5%, 2, 0, 10000)

b. = **NPER** (8%, 0, -500, 583.2)

c. = **RATE** (2, 0, -1000, 1210)

d. = **FV** (4%, 3, -1000, -500)

	PERIOD 0	PERIOD 1	PERIOD 2	PERIOD 3
PV				
PMT 1				
PMT 2				
PMT 3				

Amortization Schedule

The following shows an amortization schedule for a 30 year mortgage on an initial loan of $200,000 at 6% APR:

| F6 | | : | × | ✓ | f_x | =B6+E6 | |

◢	A	B	C	D	E	F	G
1	Principal:	$ 150,000					
2	APR:	6.00%					
3	Term:	30	years				
4							
5	Month	Beginning Balance	Monthly Payment	Interest Payment	Principal Payment	Ending Balance	
6	1	$150,000.00	($899.33)	$ 750.00	($149.33)	$149,850.67	
7	2	$149,850.67	($899.33)	$ 749.25	($150.07)	$149,700.60	
8	3	$149,700.60	($899.33)	$ 748.50	($150.82)	$149,549.78	
363	358	$ 2,671.22	($899.33)	$ 13.36	($885.97)	$ 1,785.25	
364	359	$ 1,785.25	($899.33)	$ 8.93	($890.40)	$ 894.85	
365	360	$ 894.85	($899.33)	$ 4.47	($894.85)	$ 0.00	
366							

Screenshots from Microsoft® Excel®. Used by permission of Microsoft Corporation.

Write down the formulas in each of the following cells:

5. **B6** (Beginning Principal

6. **C6** (Monthly Payment)

7. **D6** (Interest Payment)

8. **E6** (Principal Payment)

9. **F6** (Ending Principal)

10. **B7** (Beginning Principal)

Now fill down the formulas in row 6, **C6:F6**, into row 7. Then highlight all cells in row 7, **B7:F7**, and fill down!

How much interest do you pay over the 30 years? How does lowering the APR to 5% change this answer?

Investing

APR vs APY

You buy a $50,000 jumbo CD (Certificate of Deposit) with an APR of 2% compounded bi-annually. This is called a *6-month CD*, you are allowed to withdraw your funds at the end of any 6-month period, but not before (they promise to take good care of your money and water it every day).

1. How many periods are there in 1 year?

2. What is the periodic rate?

3. How much interest do you gain in the first period?

4. How much money will you have at the end of 1 year?

5. What is the APY?

Bonds (Fixed Income)

The company you work for is selling $10,000 bonds at 8% for 5 years. This means you will get *simple interest* every 6 months for the 5 year term. You buy 1 bond.

6. What is the face value?

7. What is the coupon rate?

8. What is the periodic rate?

9. How much will the company pay you every 6 months?

10. Why are bonds referred to as *fixed income* investments?

11. How much will the company pay you over the entire term (including final payment of face value)?

12. Use the geometric mean to determine the *average annual rate of return* for this investment over the 5 years.

Bond Risk

You buy a $10,000 bond at 4% for 15 years but need to sell the bond after only 6 years due to unexpected events in your life.

13. How much would someone pay for this bond if interest rates have risen to 6%?

14. Use the geometric mean to determine the average annual rate of return for this investment over the 9 years they own the bond.

15. How much would someone pay for this bond if interest rates had dropped to 3% instead of rising to 6%?

16. Looking at the following chart of the Federal Funds Rate below, which direction will the rates be heading in 2014?

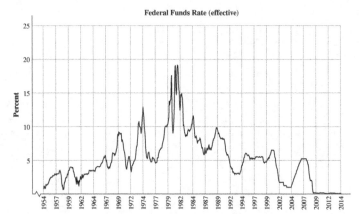

Data from: Board of Governors of the Federal Reserve System, Economic Research & Data, Selected Interest Rates (Daily) - H.15, http://www.federalreserve.gov/releases/h15/data.htm

In the United States, the **federal funds rate** is the interest rate at which depository institutions actively trade balances held at the Federal Reserve, called federal funds, with each other, usually overnight, on an uncollateralized basis. Institutions with surplus balances in their accounts lend those balances to institutions in need of larger balances.

Stocks

A friend is starting a website company and is selling 10,000 shares at $15 each; you buy 1,000 shares. At the end of the year the company makes a profit of $12,000, and decides to reinvest half into the business and payout the rest as *dividends* to the owners. Determine the following statistics:

17. *Market capitalization***:**

18. *Share price***:**

19. *Earnings per share (EPS):*

20. *Price to Earnings (PE ratio):*

21. *Dividends (per share):*

22. *Dividends (as % of share price):*

23. Someone offers to buy your shares from you for $20 a share. Do you sell?

Logic

All of the functions studied to this point have had a single output for a given set of inputs. In this chapter we now look at situations where multiple outputs are possible. Note that this seemingly violates the traditional definition of a function!

To do this we will use the **IF** and **VLOOKUP** functions.

SUM	▼	:	✕ ✓ *fx*	=IF(

▲	A	B	C	D	E	F
1	Zip Shipping					
2	Input Order Total:	$132.56				
3	Output Shipping:	=IF(				
4		IF(**logical_test**, [value_if_true], [value_if_false])				
5						
6						

Screenshots from Microsoft® Excel®. Used by permission of Microsoft Corporation.

In this example there are TWO POSSIBILITIES for the shipping amount. **IF** the order is < $100 **THEN** shipping is $7.99 **ELSE** it is free! Notice the syntax of the IF function follows this IF-THEN-ELSE format:

$$= IF(\ B2 \ < \ 100, \ 7.99, \ 0 \ ).$$

The commas are read as THEN and ELSE.

1. What function would you type in if shipping is 10% of the order, for orders less than $100, and 5% of the order for all other orders?

= IF(,   ,   )

We can also handle situations with more than two possibilities by NESTING IF functions. Assume there are three possibilities: 10% for $< \$100$, 5% for $< \$250$, and free for all others:

$$= IF(B2 < 100, 10\% * B2, IF(\underline{}, \underline{}, \underline{})).$$

2. Fill in the blanks in the previous IF function.

If there are many possibilities the nesting of IF functions gets out of hand! The **VLOOKUP** function is a better way to deal with these situations ☺, and has 3 arguments:

$$= VLOOKUP(value, table, column \#)$$

	A	B	C
1		**Zip Shipping**	
2	Bins	Cutoffs	Output %
3	$0-$99.99	$ -	10.0%
4	$100-$249.99	$ 100.00	5.0%
5	$250-$499.99	$ 250.00	4.0%
6	$500-$999.99	$ 500.00	2.5%
7	$1000-	$ 1,000.00	0.0%
8			
9	Input Order Total:	$ 874.00	
10	Shipping Percentage:	2.5%	
11	Shipping costs:	$ 21.85	
12			

In this situation we have 5 possibilities for the %. Instead of using nested IF functions we represent the possibilities in a **TABLE** in cells, **B3:C7**. This table consists of two columns, #1 and #2. Basically we will LOOKUP the **VALUE** for the order total **B9**, in the **TABLE,** and then output the appropriate % from **COLUMN # 2**:

$$=VLOOKUP(B9, B3:C7, 2)$$

In this case with **B9** = $874.00 the output of the **VLOOKUP** function will be 2.5%. Note that we use the left endpoints of the bins for the cutoff values in the table.

3. What are the outputs for the following?
=VLOOKUP(**65**, B3:C7, 2)

    b. =VLOOKUP(650, B3:C7, **1**)

    c. =VLOOKUP(6500, B3:**C6**, 2)

Lastly we will study the **RAND** function, which generates a random number between 0 and 1.

Typing in **=RAND()** results in a random number between 0 and 1, for example: 0.527834251. Note that **RAND** has no arguments but the parentheses are required, and that **IF** requires quotation marks around text in the second and third arguments.

4. What does the following **IF** function do: = IF(RAND() < 0.5, "Heads", "Tails") ?

5. Write down an **IF** function which flips an unfair coin with heads coming up 75% of the time.

6. Write down an IF function that returns the number 1, a third of the time, the number 2, a third of the time and the number 3, a third of the time.

7. Create a spreadsheet that simulates rolling a die using the VLOOKUP and RAND functions. Write down all the formulas in cells **B3:C8**, and **C10**.

	A	B	C	D
1	Random #:	0.207697		
2		TABLE		
3				
4				
5				
6				
7				
8				
9				
10	Die Roll:			

IPO Investing

Your friend who graduated with an underwater synchronized dance degree just landed a job at a brokerage house (answering phones) where they overheard a hot stock tip. By investing in ultra-risky initial public offerings (IPOs) each week you either can earn 80% or lose 60% (completely random 50-50 chance of which occurs), which means you should average a 10% gain! Right?! RIGHT?!

1. Assuming you invest $10,000 and then average a 10% gain each week for 52 weeks, compute how much your investment is now worth.

2. Consider the following spreadsheet shown which randomly assigns one of the outcomes, 80% or -60%, for each week in 1 year using the IF function and the RAND function: IF RAND()<0.5 THEN 80% ELSE -60%.

	A	B	C	D	E	F	G	H
1	Week	Balance	Return	End Balance		Balance at the end of the year:		
2	1	$ 10,000.00	-60%	$ 4,000.00		$ 1.95		
3	2	$ 4,000.00	-60%	$ 1,600.00				
4	3	$ 1,600.00	80%	$ 2,880.00				
5	4	$ 2,880.00	-60%	$ 1,152.00				
6	5	$ 1,152.00	-60%	$ 460.80				
7	6	$ 460.80	-60%	$ 184.32				
8	7	$ 184.32	80%	$ 331.78				
9	8	$ 331.78	80%	$ 597.20				
10	9	$ 597.20	80%	$ 1,074.95				
11	10	$ 1,074.95	-60%	$ 429.98				
12	11	$ 429.98	80%	$ 773.97				
13	12	$ 773.97	-60%	$ 309.59				

    i. What formula is entered in cell C2?

ii. What formula is entered in cell D2?

iii. What formula is entered in cell B3?

iv. What formula is entered in cell F2?

3. How does the end of year balance shown compare to your result in **part 1**?

4. If we keep track of the ending balance for 10 different scenarios we get the results shown. Average these 10 ending balances.

    i. Scenario 1: $1.95
    ii. Scenario 2: $0.00
    iii. Scenario 3: $177.94
    iv. Scenario 4: $1.95
    v. Scenario 5: $800.75
    vi. Scenario 6: $0.10
    vii. Scenario 7: $0.00
    viii. Scenario 8: $1.95
    ix. Scenario 9: $800.75
    Scenario 10: $8.79

5. Is this average close to what you computed in the first part of this problem? Why not!?

6. Compute the geometric mean of 80% and -60%. Use this average rate of return to re-compute the worth of your investment after 52 weeks. Is this closer to your average of 10 ending balances?

Know Your Chances

The table below gives the risk of dying from various diseases for 55 year old men and women by smoking status. These numbers are rates: deaths per 1,000 of each cohort over the next 10 years.

Table 11.1: Death Rates

Age	Sex	Smoking Status	Vascular Disease		Cancer					Lung Disease	Accidents	All Causes
			Heart Attack	Stroke	Lung	Breast	Colon	Prostate	Ovarian	COPD		
55	M	Never Smoked	19	3	1		3	2		1	5	74
55	M	Smoker	41	7	34		3	1		7	4	178
55	F	Never Smoked	8	2	2	6	2		2	1	2	55
55	F	Smoker	20	6	26	5	2		2	9	2	110

Data From: Steve Woloshin, Lisa Schwartz, and H. Gilbert Welch, "The Risk of Death by Age, Sex, and Smoking Status in the U.S.: Putting Health Risks in Context," *Journal of the National Cancer Institute* 100 (2008): 845–853[1]

Is smoking bad for you? How do you know?

1. Who is more likely to die in the next 10 years from a heart attack: a 55 year old man or woman?

2. By what factor does smoking increase the chance of a 55 year old man/woman dying from lung cancer?

3. Does smoking cause heart attacks or minimize the chance of getting breast cancer?

---

[1] http://www.ncbi.nlm.nih.gov/pmc/articles/PMC3298961/

4. Should you smoke?

Does human activity influence the climate? How do you know?

The Intergovernmental Panel on Climate Change (IPCC) issued their fifth assessment report (AR5[2]) in 2014 and concluded that human greenhouse gas emissions "have been detected throughout the climate system and are *extremely likely* (95–100% certain) to have been the dominant cause of the observed warming since the mid-twentieth century." Note that the figure below shows observed data in line with scientists' predictions. Surface temperature is projected to rise over the 21st century under all assessed emission scenarios. It is *very likely* (90–100% certain) that heat waves will occur more often and last longer, and that extreme precipitation events will become more intense and frequent in many regions. The ocean will continue to warm and acidify, and global mean sea level to rise.

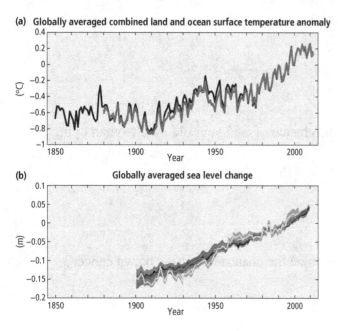

(a) Globally averaged combined land and ocean surface temperature anomaly

(b) Globally averaged sea level change

---

[2] http://www.ipcc.ch/pdf/assessment-report/ar5/syr/SYR_AR5_FINAL_full.pdf

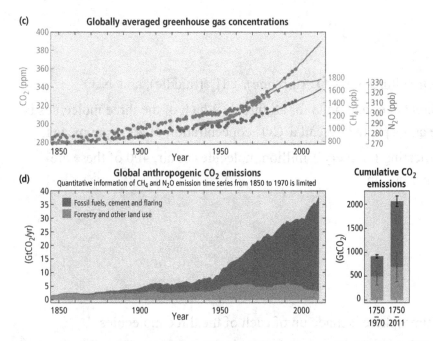

(c) **Globally averaged greenhouse gas concentrations**

(d) **Global anthropogenic CO₂ emissions**
Quantitative information of CH₄ and N₂O emission time series from 1850 to 1970 is limited

Fossil fuels, cement and flaring
Forestry and other land use

Cumulative CO₂ emissions

Source: Text from Climate Change 2014: Synthesis Report. Contribution of Working Groups I, II and III to the Fifth Assessment Report of the Intergovernmental Panel on Climate Change. [Core Writing Team, Pachauri, R.K. and Meyer, L. (eds.)], pp 3 IPCC, Geneva, Switzerland.

5. In **Figure (a)** interpret the value for 1850 and 2010.

6. In **Figure (b)** the data looks linear. Sketch in a straight line and find the equation.

7. Interpret the slope and y-intercept of your line.

8. In **Figure (c)** there are 3 trendlines for CO_2 (top line), CH_4 (middle), and N_2O (bottom). There are different axes and units for all three! Which of the three molecules is most concentrated in the atmosphere? Recall a CO_2 concentration of 400 ppm means 400 parts per million, indicating for every 1 million molecules of air, 400 of these are CO_2. Also ppb is parts per billion.

9. What percentage of the atmosphere is made up of each of the three molecules?

10. Given these are such small percentages, why is there cause for concern about the indicated concentrations?

11. In **Figure (d)** how does the graphic on the left differ from the cumulative emissions graphic on the right?

12. There is uncertainty in the cumulative CO_2 emissions from 1750 to 2011. What do scientists are the smallest and largest cumulative CO_2 emissions over this time period?

13. What is the estimated cumulative CO_2 emissions from 1970 to 2011?

14. Find the average cumulative CO_2 emissions per year over the two different time periods: 1750–1970 and 1970–2011.

15. We have seen above how bad smoking is for us. Does it seem reasonable to believe that human caused greenhouse gas emissions have no effect on the climate system?

Know Your Children

Computing probabilities involves careful counting as the following definition spells out:

Definition: The **probability** of an *event* occurring, $P(Event)$, is the ratio of the number of ways this event can occur to the total number of *outcomes* possible. The collection of all possible outcomes is called the *sample space*, and the event is considered to be a subset of the sample space. Thus the probability ratio can always be represented as a fraction/decimal between zero and one.

- $0 \leq P(Event) \leq 1$

- $P(Event) = 0$ implies it will not occur, zero chance.

- $P(Event) = 1$ implies it will occur, guaranteed 100% certain.

Compute the Probability of a Daughter

You meet with a friend who has two children. What is the probability her second child is a girl? Assume the probability of having a boy or girl is equally likely (50%).

1. What is the sample space for all possible ways she can have 2 children?

2. List the outcomes for the event, *2nd child girl*.

3. Now compute the probability, $P(2nd\ child\ girl)$.

She mentions one of her children is a boy. Given this new information what is the probability that her second child is a girl?

4. How does the new information change the sample space?

5. Now compute the probability, $\mathbf{P}(\textit{2nd child girl})$.

Lastly she tells you the previously mentioned boy was born on a Tuesday. Given this new information what is the probability that her second child is a girl?

6. How does the new information change the sample space?

7. Now compute the probability, $\mathbf{P}(\textit{2nd child girl})$.

Definition: Given a series of k choices to be made, each with a different number of options: n_1, n_2, ..., n_k, then the total number of possible ways to choose 1 option from each of the k choices is the product:

$n_1 \times n_2 \times \cdots \times n_k$. This is called the **Fundamental Principle of Counting**.

8. You have 2 shirts, 3 shorts, and 3 hats to choose from to make an outfit. How many different outfits are possible?

9. Draw a tree diagram illustrating the possibilities.

Definition: A **permutation** is a selection of r items from a group of n, with no item being selected twice and the order items are selected is counted differently (abc is a different permutation than bca). The total number of permutations is given by:

$$_nP_r = n\times(n-1)\times(n-2)\times\cdots\times(n-r+1) = \frac{n!}{(n-r)!}$$

A **combination** is a selection of r items from a group of n, with no item being selected twice and the order items are selected is NOT counted differently (abc is the same combination as bca). The total number of combinations is given by:

$$_nC_r = \binom{n}{r} = \frac{n\,Pr}{r!} = \frac{n!}{(n-r)!\cdot r!}$$

10. Given 12 people in a race, how many ways can they finish first and second?

11. Given 12 people in a class, how many ways can we choose a committee of 4?

Definition: The **conditional probability** of *A given B* is written $P(A \mid B)$ and is the ratio of the number of outcomes occurring in both events *A* and *B*, to the number of total outcomes in *B*.

Given the following table:

	MALE	FEMALE	TOTALS
Smoker	130	75	**205**
Non-smoker	250	215	**465**
Totals	**380**	**290**	

12. Compute the probabilities: **P**(*Smoker* | *Male*) and **P**(*Female* | *Non-smoker*).

It's the Law!

Computing probabilities involves careful use of the 3 laws of probability:

Definition: The **1st Law of Probability** says that the probability of event A and event B both occurring is:

$$P(A \text{ AND } B) = P(A) \cdot P(B \mid A)$$

If A and B are *independent* events then $P(B \mid A) = P(B)$ and the formula is:

$$P(A \text{ AND } B) = P(A) \cdot P(B)$$

Flipping two coins are *independent* events, the probability of the second flip does not depend on the first. Choosing two committee members are *dependent* events, the probability of choosing the second member depends on the first choice.

Apply the 1st Law of Probability

A sub-committee of 2 people will be chosen from a group of 6: {Sue, Ann, Jill, Bob, Jim, Joe}. The first 3 listed are women and the last 3 listed are men.

1. What is the sample space for this problem?

2. Compute the **probability** the sub-committee contains 2 women using the 1st law.

Definition: The **2nd Law of Probability** says that the probability of event A or event B occurring is:

$$P(A \text{ OR } B) = P(A) + P(B) - P(A \text{ AND } B)$$
$$P(A \cup B) = P(A) + P(B) - P(A \cap B)$$

If A and B are *disjoint* events then $P(A \text{ AND } B) = 0$ and the formula is:

$P(A \text{ OR } B) = P(A) + P(B)$.

The probability of event A or event B occurring means A or B or both, and is the same as the probability of *at least one* of event A or event B occurring.

Apply the 2nd Law of Probability

A sub-committee of 2 people will be chosen from a group of 6: {Sue, Ann, Jill, Bob, Jim, Joe}. The first 3 listed are women and the last 3 listed are men.

3.   What is the **probability** that at least one man is chosen?

4. Compute the **probability** the sub-committee contains 1 man and 1 woman using the 1st law and the 2nd law.

Definition: The **3rd Law of Probability** says that the probability of event A is one minus the probability of its complement, **NOT** A:

$$\mathbf{P}(A) = 1 - \mathbf{P}(\text{NOT } A)$$

$$\mathbf{P}(A \textbf{ OR } B) = 1 - \mathbf{P}(\text{NOT } A \textbf{ AND } \text{NOT } B)$$

For *independent* events we can apply the **1st Law** to the last formula and get the *Golden Rule of Probability*:

$$\mathbf{P}(At\ Least\ One\ Independent\ Event(A,\ B,\ C,...)\ Occuring) = 1 - \mathbf{P}(\text{NOT } A) \times \mathbf{P}(\text{NOT } B) \times$$
$$\mathbf{P}(\text{NOT } C) \times ...$$

Apply all Three Laws

5. Flip two coins and determine the probability of getting two heads $(2H)$ and the probability of getting at least one head $(At\ least\ 1H)$.

Apply the Golden Rule

6. You are applying to 5 colleges and believe you have a small 10% **chance** of getting into any one of them. What is the **probability** of getting into at least one of the schools?

From Conditional Probability to Bayes

In an experiment to determine if a skin cream helps a rash, the following data is collected:

	Rash Got Better	Rash Got Worse	Totals
Did use cream	223	75	298
Did not use cream	107	21	128
Totals	330	96	426

1. Does the skin cream help the rash? How do you know?

2. Draw a Venn Diagram of this information. Which representation is easier for you to process?

3. Compute and compare the probability of someone getting better given they used the cream, $\mathbf{P}(Got\ Better\,|\,Used\ Cream) = \mathbf{P}(GB\,|\,UC)$, and the probability of someone having used the cream given they got better, $\mathbf{P}(Used\ Cream\,|\,Got\ Better) = \mathbf{P}(UC\,|\,GB)$.

4. Which representation of the information, two-way table or Venn diagram, was more helpful to you in computing the probabilities?

5. Do you know how these two probabilities are related?

Starting with the formula for conditional probability:

$$P(UC \mid GB) = \frac{P(UC \text{ AND } GB)}{P(GB)} = \frac{P(UC) \cdot P(GB \mid UC)}{P(GB)} = \frac{P(UC)}{P(GB)} \cdot P(GB \mid UC)$$

Now the people who got better (GB) either used cream (UC) or did not (NoC) so:

$$P(GB) = P(GB \cap UC) + P(GB \cap NoC)$$

$$P(GB) = P(GB \mid UC) \cdot P(UC) + P(GB \mid NoC) \cdot P(NoC)$$

Substituting in for the denominator above gives us Bayes' Formula:

$$P(UC \mid GB) = \frac{P(UC) \cdot P(GB \mid UC)}{P(GB \mid UC) \cdot P(UC) + P(GB \mid NoC) \cdot P(NoC)}$$

6. Recompute $P(UC \mid GB)$ using Bayes' Formula above.

Now see if you can follow how the following Bayes Table was constructed given the data that someone got better:

Hypotheses (Hᵢ)	Priors $P(H_i)$	Likelihoods $P(D \mid H_i)$	Products $P(H_i) \cdot P(D \mid H_i)$	Posteriors $P(H_i \mid D)$
Did use cream	298/426 = 70.0%	223/298	223/426	223/330 = 67.6%
Did not use cream	128/426 = 30.0%	107/128	107/426	107/330 = 32.4%
Totals	100%		330/426	100%

7. How does the table indicate whether or not the cream works?

Apply Bayes' Formula

Your father is 55 years old and has just had a heart attack. He says he is not smoking but you think there is a 20% chance he has been smoking. Given that he has had a heart attack what is the new probability that he is smoking? Our two hypotheses are that he is smoking (S) or not smoking (NS) and Table 11.1 from the text gives us the likelihood of a heart attack (A) in both cases: $P(A|S) = 4.1\%$ and $P(A|NS) = 1.9\%$.

8. Fill in a Bayes Table and compute the new probability of smoking given the new data of a heart attack:

HYPOTHESES (H_i)	PRIORS $P(H_i)$	LIKELIHOODS $P(D|H_i)$	PRODUCTS $P(H_i) \cdot P(D|H_i)$	POSTERIORS $P(H_i|D)$
Smoking				
Non Smoking				
Totals	100%			100%

Apply Bayes' Formula with 3 Hypotheses

Three siblings (Galen, Mei, and Beau) are equally likely to eat an apple, but the probability they leave the core on the counter differs: $P(Core|Galen) = 1/5$, $P(Core|Mei) = 3/5$, and $P(Core|Beau) = 4/5$. If we find a core on the counter, what is the new probability that each sibling was the culprit? Our three hypotheses are that each sibling ate the apple and left the core on the counter.

9. Fill in a Bayes Table and compute the new probabilities of each sibling eating the apple given the new data of a core:

HYPOTHESES (H_i)	PRIORS $P(H_i)$	LIKELIHOODS $P(D\|H_i)$	PRODUCTS $P(H_i) \cdot P(D\|H_i)$	POSTERIORS $P(H_i\|D)$
Galen				
Mei				
Beau				
Totals	100%			100%

Solve the Monty Hall Problem

The Monty Hall Problem is a classic probability conundrum. It has fooled experts with advanced degrees in statistics and led to public embarrassment as these experts attacked each other in print. The problem states that you are on a game show and must choose 1 out of 3 doors. There is a car behind one of the doors. After you have selected your door, the host, Monty Hall, opens another door that does not have the car (he knows where the car is) and asks if you want to change your door. Should you switch? Our three hypotheses are that the car is equally likely to be behind one of the 3 doors. Let's assume you pick door A and Monty opens door B, should you switch to door C?

10. Compute the likelihood that Monty opens door B given the data that the car is behind door A.

11. Compute the likelihood that Monty opens door B given the data that the car is behind door B.

12. Compute the likelihood that Monty opens door B given the data that the car is behind door C and fill in the table.

HYPOTHESES (H_i)	PRIORS $P(H_i)$	LIKELIHOODS $P(D|H_i)$	PRODUCTS $P(H_i) \cdot P(D|H_i)$	POSTERIORS $P(H_i|D)$
Door A				
Door B				
Door C				
Totals	100%			100%

One last application of Bayes' Theorem to making informed health decisions. There exist many tests for diseases. A false positive indicates you test positive for the disease (indicating sick) but don't actually have it, while a false negative indicates you test negative for the disease (indicating healthy) but actually have it.

Predict Illness

A certain disease occurs in 5 out of 1,000 people. A test has a false positive rate of 3% and a false negative rate of 1%. What is the probability that a person who tests positive actually has the disease, $\mathbf{P}(Sick\,|+)$?

13. What is the probability that a person actually has the disease, $\mathbf{P}(Sick)$?

14. What is the probability that a person does not have the disease, $\mathbf{P}(Well)$?

15. What is the probability that a person tests positive given they actually have the disease, $\mathbf{P}(+|\,Sick)$?

16. What is the probability that a person tests positive given they actually have the disease, $P(+|Well)$?

17. Now fill in the Bayes table to determine the probability that a person who tests positive actually has the disease, $P(Sick|+)$:

HYPOTHESES (H_i)	PRIORS $P(H_i)$	LIKELIHOODS $P(D|H_i)$	PRODUCTS $P(H_i) \cdot P(D|H_i)$	POSTERIORS $P(H_i|D)$
Sick				
Well				
Totals	100%			100%

18. Now assume the population consists of 100,000 people and create a two-way table like in the skin cream experiment and determine the probability that a person who tests positive actually has the disease, $P(Sick|+)$. Is this easier than the Bayes Table?